Hello! Middle School Mathletes !!

The Preface

Let me guess - you will be taking a CML test or an IMLEM test or an MOEMS(Div M) test or the AMC-8 test tomorrow or in a week. Don't worry if the acronyms appear to be unfamiliar. Those were unfamiliar to me until my middle schoolers brought home fliers with these words. Let's just say you will be taking a competitive math test under the influence of self, teacher, parent, grade, peer or a silly challenge. That you are looking at this book is a proof that inside you is working a math genius who dreads at the thought of doing tons of repetitions of the same problem with just numbers changed. You are looking at the right book.

This book presents 109 problems categorized into 7 chapters, each chapter corresponding to a major topic. These problems can be solved using the fundamental concepts learned in the middle school. Some of the problems may seem easier for the students attending courses at the honors level, sometimes referred to as gifted level, nonetheless they can be solved using fundamental basic skills. The underlying principle for a lot of problems has been to guide the students to linking of multiple concepts, where I have seen a lot of middle schooler mathletes to stumble.

I would like to thank Ms. Tripti Agarwal and Mr. Biswarup Chattaraj, who are two accomplished engineers for reviewing the manuscript and tendering valuable suggestions.

I enjoyed creating these problems. My greatest reward will be the joy the budding mathletes will derive when they solve these problems and strengthen their mathematical thinking process.

Your feedback and comments are always welcome.

S. Mandal

Index

PROBLEMS

P1. Exponents

1.1. Solve for x and y if $5 \cdot 50^y = 2 \cdot 4^x \cdot 5^{2x}$

1.2. Solve for x and y, given $10^y = 8 \cdot 4^x \cdot 5^{2x}$

1.3. Solve for x and y, given $\sqrt[x]{10^y \cdot 25} = 500$

1.4. Which of the following numbers are perfect cubes?

a. 4^{13} b. 8^{13} c. 20^{27} d. $4^{100} \times 2^{100}$

1.5. Simplify $\sqrt[5]{16^{2n}}\sqrt[5]{64^{2n}}$

1.6. Find k if $\sqrt[3]{x^5} \times \sqrt[3]{x^4} \times x^k = x^2$

1.7. For what value of x, $4^{2/x} = \frac{1}{\sqrt[4]{8}}$

1.8. Solve for x: $4 \cdot 4^2 \cdot 4^3 \cdot 4^4 \cdot \ldots \cdot 4^{24} = 8^x$

1.9. Given $\sqrt[x]{q^y} \times \sqrt[y]{q^x} = q^{\frac{10}{xy}}$, where x and y are not zeros and q is a real number > 0, Find the value of $4(x^2 + y^2)$.

1.10. Solve for x and y if $\sqrt{8 \cdot 3} \cdot \sqrt[4]{27 \cdot 2} = 8^x \cdot 81^y$

1.11. What is the unit digit of $3^{1042} \cdot 4^{99}$?

P2. Using Standard Formulas of Square and Sums

2.1. Find $p^2 + \frac{1}{p^2}$ if $p + 1/p = 10$

2.2. Find $p^2 + \frac{1}{p^2}$ if $p - 1/p = 9$

2.3. Find $p^3 - \frac{1}{p^3}$ if $p - 1/p = 4$

2.4. Find $p^3 + \frac{1}{p^3}$ if $p + 1/p = 4$

2.5. Simplify: $\frac{6.3\times6.3\times6.3 - 2.3\times2.3\times2.3}{6.3\times6.3 + 6.3\times2.3 + 2.3\times2.3}$

2.6. Simplify: $\frac{499\times499\times499 + 201\times201\times201}{499\times499 - 499\times201 + 201\times201}$

2.7. Given that the sum of squares of two positive numbers is 125 and their product 37.5, what is the difference of the squares of these two numbers?

2.8. If the difference between squares of two consecutive natural numbers is 33, what are the numbers?

2.9. Squares of two numbers add up to 200 and numbers differ by 40. What is 3 times the product of these two numbers?

2.10. Given

$$2x^2 - 2xy + y^2 + 3x + 2.25 = 0, \ Find\ x\ and\ y.$$

2.11. $a^2 + b^2 + c^2 - ab - bc - ca = 0\ and\ a = 4.$

$Find\ a + 2b + 3c.$

P3. Geometry

3.1. Area of an equilateral triangle is $16\sqrt{3}$ cm^2. What is the length of one of its sides?

3.2. What is the sum of all the interior angles of an n-sided polygon? Hence find the measure of each angle of a regular dodecagon.

3.3. An isosceles triangle has its base = b cm, and one of its equal sides l cm long.Express the area of the triangle in terms of l and b.

3.4. A square ABCD with area = $13cm^2$ is inscribed in a bigger square PQRS with area = $18cm^2$. Vertex A lies on the side PQ such that PA=a cm and QA = b cm. Find ab.

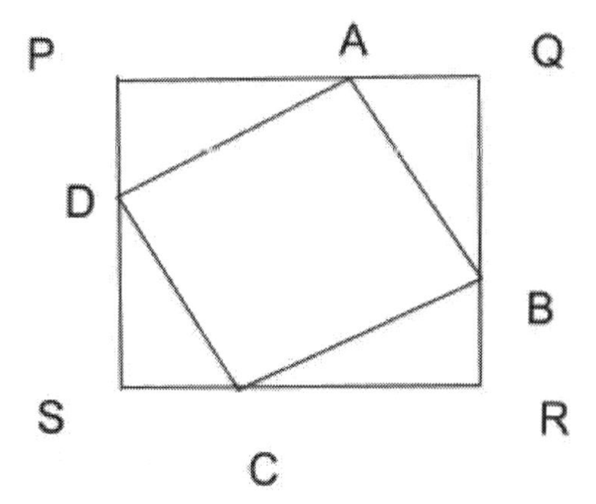

3.5. Two diagonals of the rectangle ABCD intersect at E. A straight line through E intersects AC and BD at P and Q respectively. Prove: ΔAEP is congruent to ΔDEQ. Hence find the percent area of the rectangle covered by the shaded region.

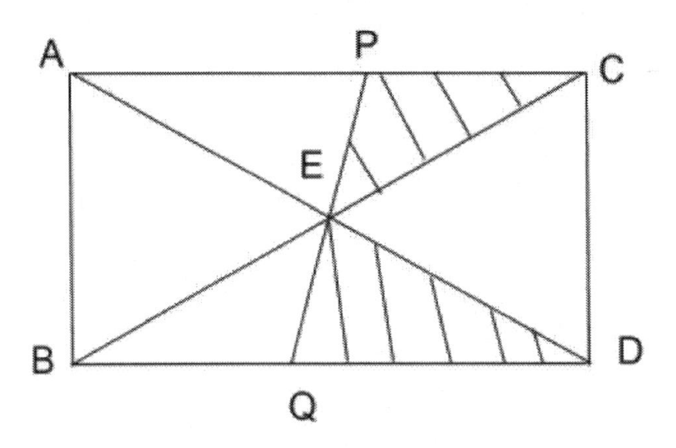

3.6. In rectangle ABCD a line intersects sides AB and BC at E and F respectively. Also given is that AE : BE = 2: 1 and CF : BF = 2:1.Compute areas of $\Delta AED,\ \Delta DFC\ and\ \Delta BEF$ as fraction of area of the rectangle. Now find the area of the shaded region as fraction of the area of the rectangle.

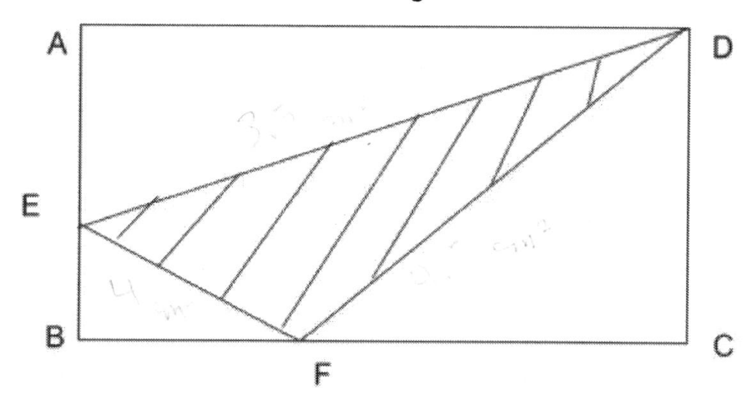

3.7. The areas in square cm of three triangular partitions of a rectangle ABCD are as shown in the figure, What is the area of the rectangle?

[Hints: express AE and FC in terms of x and y using area of two triangular partitions, and then express area of the remaining triangle]

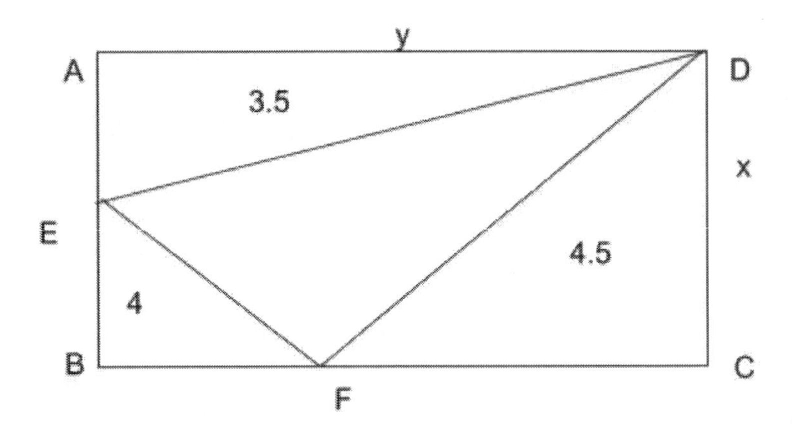

3.8. In Triangle ABC, the bisector of ∠BAC intersects side BC at P. Side BA is extended by the length of side AC to point Q, and Q is joined to C. if AB : AC = 2:1 and, and AP = 10cm. Find the length of CQ.

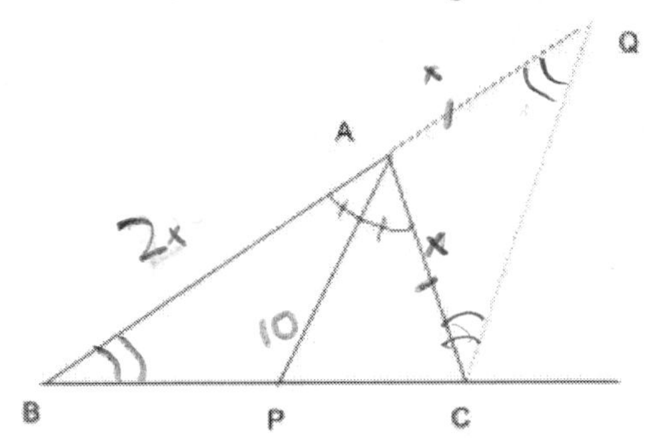

3.9. ABCD is a square and ΔAPD is a triangle right angled at P. If AP=6cm and PD= 8cm, find the areas of ΔAPB and ΔCPD.

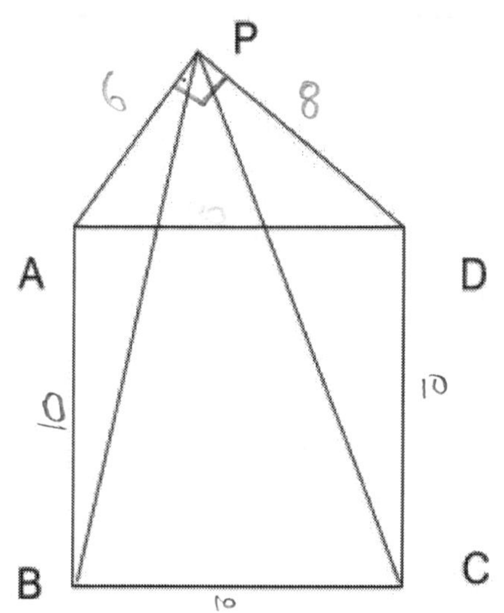

3.10. In triangle ΔABC, ∠B = 90°. BP is perpendicular to AC at point P. If length of BC = 5 cm and length of BA = 12cm, Find BP, PC and AP.

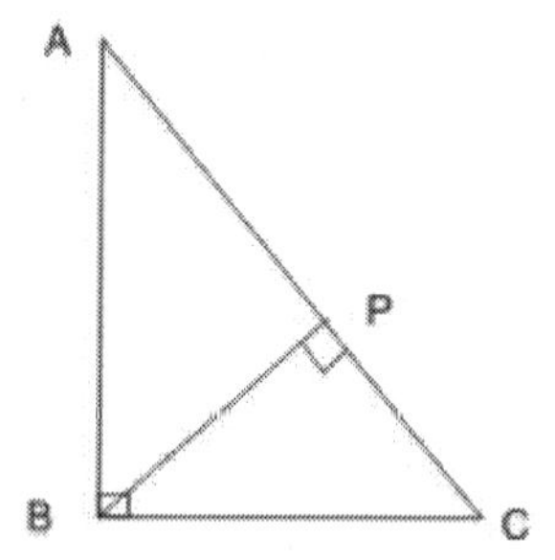

3.11. A circle with centre O is divided into 15 equal arcs and marked A through O in the clockwise order. First arc is between A and B, 2nd between B and C, and so - last arc is between O and A. A, C, D, G, H and L are joined to O with line segments. What are the measures of ∠AOC, ∠DOG and ∠HOL? A and C are joined by a straight line segment AC. Similarly H and L are joined by a straight line segment HL. Also find angles $\angle ACO$ and $\angle HLO$.

3.12. Area of a right triangle is 10 sq cm and its hypotenuse is $\sqrt{104}$ cm. What is the difference of lengths of other two sides ?

3.13. A circle is inscribed in an equilateral triangle, and a larger circle circumscribes the triangle. Find the ratio of circumferences of the two circles.

3.14. An equilateral triangle is inscribed in a circle, and the circle is inscribed in a bigger equilateral triangle. If the area of the region between the smaller triangle and the circle is $(\frac{\pi}{3} - \frac{\sqrt{3}}{4})\,16$ square units, what is the area between the larger triangle and the circle?

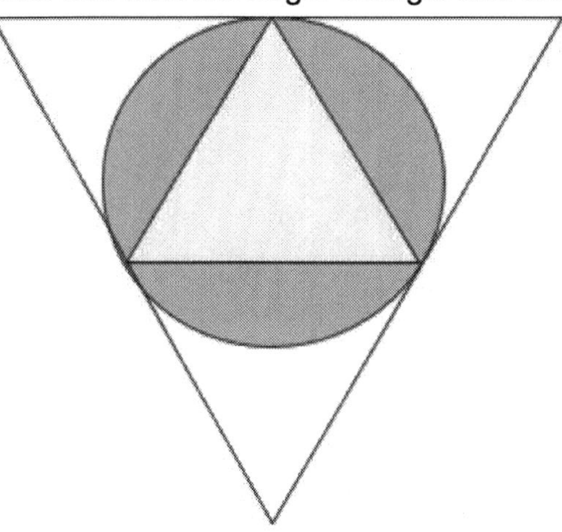

3.15. From a point P on the circle $\odot O$ with radius 10 cm is drawn a line segment $\overline{PN}$ perpendicular to the diameter AB at N. If ON is 5cm, what is the length of $\overline{PN}$?

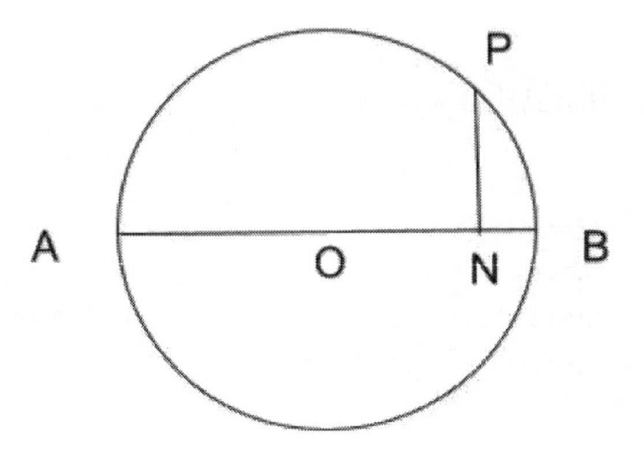

3.16. A circle with centre (5,0) is tangent to the line y=2x. What is the area of the circle?

P4. Functions

4.1. Find the quadratic function, if 3 and 10 are its zeroes and the maximum value of the function in 5.

4.2. Find a cubic polynomial if it passes through (0, 0) and f(1) =1 , f(2)= 2 and f(3) = 4.

4.3. If $f(x) = x^3 + 4x + 2$, find f(2y+1)

4.4. If $f\left(\frac{3+x}{2}\right) = 3x$, find f(x)

4.5. If $f\left(\frac{2+3x}{5}\right) = x^2 + 5x + 4$, find f(x)

4.6. Find the domain and range of $f(x) = \sqrt{x^2 + x}$

4.7. Find Domain and Ranges:

a) $f(x)=3^{-x}$	b) $f(x) = \frac{2}{\lvert x-3 \rvert}$	c) $f(x) = -(x-4)^{-2}$
d) $y = \lvert x-3 \rvert + 4$	e) $f(x) = x^3 + x$	f) $f(x) = (x-2)^2 + 6$
g) $y = \sqrt{x-3}$	h) $y = \sqrt{\lvert x-3 \rvert}$	i) $y = 3x^{\frac{1}{4}}$

4.8. A linear function passes through (3, 4) and (7, -8). What is its y intercept and slope?

4.9. What is the y intercept of the function $y=3x^3 + 5x^2 + 3x + 5$. Find the value of x for which f(x)=0.

4.10. If $f(n) = f(n-1) + n$ and $f(1) = 1$, find explicit expression of $f(n)$.

4.11. If the explicit form of an arithmetic progression is given by $a_n = 5n - 11$, find a_1 and d.

4.12. If the explicit form of an arithmetic progression is given by $a_{n-1} = 7n + 11$, find a_1 and d.

4.13. A function y= f(x) has y-intercept 3 and a minimum value at (4, -6). Find y intercept and minima of $y = f(x)+4$.

4.14. If a function $y=-4(x-3)^2+2$ has maxima at (3, 2), where would be the maximum point of $y=-4(x-2)^2 -4$

4.15. If f(x) has zeros at x= 4, 7 and 9, where would be zeroes of f(x-3).

4.16. An exponential function has the following table:

x	y
0	2
1	6
3	s
t	$\frac{2}{3}$

What are the values of s and t?

4.17. If an arithmetic progression has has $a_{21} = 25$ and $a_{43} = 58$. Write the explicit formula for the progression.

4.18. Write an equation of a line that is perpendicular to line $3x + 4y = 20$, and passes through the point (-10,2).

4.19. Write the recursive formula in function notation for the sequence, $a_n = \frac{1}{4} - \frac{7}{20}(n-2)$

4.20. In an arithmetic progression, for a given k, $a_k = 25 \text{ and } a_{k-3} = 19 \text{ and } a_3 = 5$. What is the value of k?

4.21. Let $S_n = a_1 + a_2 + a_3 + a_4 + \ldots + a_n$, where $a_1, a_2, a_3, \ldots, a_n$ are in arithmetic progression with common difference d. Show that

$a_k + a_{n-k+1} = 2a_1 + (n-1)d = a_1 + a_n$ and

$S_n = \frac{n}{2}(a_1 + a_n)$

4.22. In the sequence 2, 9 , 16, 23, … which term is 373?

4.23. If sum of 11 consecutive integers is 121, what is the starting number?

4.24. If sum of 23 consecutive odd numbers is 2047, what is the starting odd number?

4.25. For what value of n, $81^n = 3^2 \cdot 3^4 \cdot 3^6 \cdot \ldots \cdot 3^{2n}$?

P5. Speed, Time, Work

5.1. How long does a 200m long train take to cross a pole, if the train is travelling at speed = 60 kmph?

5.2. How long does a train 200m long take a cross a bridge 1.3km long if it is travelling at 90 km per hour?

5.3. Anne runs 25% faster than Bobby. They started at the same time from the same starting point along an oval track. When they cross each other how many laps each of them has finished?

5.4. Natalie runs 30% faster than Pollie. They started at the same time from the same starting point on a 420m oval track. When they crossed each other, what is the shortest distance between the cross point and the start line?

5.5. Rita can finish painting a wall in 12 days. After 3 days her daughter joins, and they are able to finish the rest of the painting in 6 days. How many days would the daughter take to paint the entire wall if she worked alone?

5.6. A and B can finish a piece of work in 6 days. B and C can finish same piece of work in 9 days, and A and C in 12 days. How many days will each of them take finish if they worked individually?

5.7. A water tank has an inlet, which when open fills the entire tank in 12 hrs. It has an outlet which when open empties the full tank in 20 hrs. The tank is supplying water continuously through the outlet and cannot be shut. When the tank is half emptied, the inlet was turned on. How long will it take for filling the tank to the top again?

5.8. Three faucets can fill a tub in 10 min, 15 min and 20 min respectively. All the three faucets were opened at the same time fill the empty tank. After 2 min, the first faucet was turned off. In how many more minutes will the tank be completely filled?

5.9. Two boys take a shortcut through a railway tunnel. Their worst dream comes true - when they are two thirds way through the tunnel, they see a train coming on the track. They start running in opposite directions. Both of them get out just in time. If the train's speed was 60km/hr, and both the boys were able to accelerate instantaneously to the same speed v, find v.

5.10. Two trains are approaching on same track with speeds 60 kmph and 90 kmph, and they are 1.5km apart. A dumb bee stuck between the two trains, while trying to get out, starts flying towards one train, makes a u-turn as it reaches the train. Similarly when it reaches the other train, it makes a u-turn, and it keeps doing so. What distance would the bee cover, before the trains collide, if the bee flies at 120 kmph with respect to stationary reference?

5.11. John bikes to the end of a trail at the speed of 30 kmph and comes back retracing the same trail at 20 kmph. What is the average speed of John?

5.12. JK is on a train(A) traveling at 36 kmph and sees a train(B) coming the opposite direction. Train B is traveling at 24 kmph and lengths of the trains A and B are 240m and 200m respectively. JK uses her watch to note duration of train crossing her. What was the time she noted?

5.13. JK and MJ are running at the speeds 4m/s and 7.2 km/hr speeds respectively. If they started at the same time and MJ was 200m ahead, how long would JK take to get 100m ahead of MJ?

5.14. A gun was fired twice at the interval of 21 minutes. A man running towards a gun field hears two shots at the interval of 20 min 15 seconds. What is man's speed, if speed of sound is 351 m/s?

5.15. Standing at on the side of a railway track, MJ found that two trains of same length crossed him in opposite directions in 18 sec and 12 seconds respectively. What time would the trains take to cross each other?

5.16. Speeds of the two boats A and B are 9 kmph and 8 kmph downstream respectively. If the speed of A be 7 kmph upstream, how much more time will be taken by the slower boat to cover 112 km in still water?

5.17. The Bullet train goes from station A to station B traveling at 300 kmph for the first half and at 360 kmph for the second half. While coming back it covers ⅖ of the journey at 250 kmph and rest at 390 kmph. What was the train's average speed over the to and fro journey?

5.18. April and May start running from the same start point at the same time in opposite directions around an oval track. Their speeds are 9 kmph and 12 kmph until they cross each other. From that point what speed must April run if she plans to meet May at the start point when she completes her round? Assume both are physically capable to do this.

5.19. To finish a long stretch of road work, 20 workers started working 8 hours a day. After elapsing $\frac{3}{4}$ scheduled time, it was found that $\frac{2}{5}$ of the work had been finished. How many more workers need to be absorbed so that work can be finished in planned time, working 10 hours per day?

5.20. 3 men and 5 women can finish a piece of work in $2\frac{2}{5}$ days. 6 men and 4 women can finish the same piece of work in $1\frac{7}{8}$ days. How long will 9 men and 10 women take to finish the work?

P6. Probability

6.1. A dart board is in the shape of a concentric square and a circle. Radius of the circle is same as the length of a side of the square. If the probability of missing the target is 0.2, what is the probability that a dart thrown randomly at the target lands in the shaded area?

6.2. Mrs Bugaboo tells her student bugs that there are 3 problems out of a practice set of 10 problems coming on mid-term test. BugZ chooses to practice only 7 problems from the practice set. What is the probability that he finds that all the three problems on the test were practiced by him?

6.3. From a deck of 52 cards, Ramona draws a card and it is an ace. She now draws two cards. What is the probability that both the cards are either black or Jacks?

6.4. If the probability of winning the jackpot in a casino in each visit is 0.2, what is the probability of Peter winning twice in the casino if he visited the casino 4 times?

6.5. If the probability of Romy forgetting his umbrella at home is x, and not forgetting is y, find the value of the expression, $8x^2 + 16xy + 8y^2 + 10(x + y)$

6.6. There are 4 boys and 3 girls on the editorial board of the school. What is the probability of having at least one boy on each side in a photo, if a random photo is shot for the entire team?

6.7. Six people take seats randomly in a sitting area. What is the probability that three friends find themselves sitting next to each other?

6.8. Two integer numbers are chosen at random from the set of integer numbers, { 31, 32, .., 65} What is the probability that both numbers add up to 73.

6.9. An unbiased dice is rolled twice. What is the probability that the sum total of points is 8?

6.10. There are three urns with the following ball counts :

a. U_1: 2 red, 3 blue, 2 green
b. U_2: 3 red, 4 blue, 3 green
c. U_3: 2 red. 2 blue, 2 green

If 1 ball is drawn at random from each urn, what is the probability of getting all different colored balls?

6.11. There are northbound trains at 5:00pm, 5:30pm, and 6:00pm from a rail stations. There are South bound trains at 5:10 pm, 5:45 pm and 6:10 pm from the same station. If Abbey arrives randomly between 5:00 pm and 6:00pm, what is the probability that the first train he sees is a southbound train?

6.12. Two numbers are picked at random from the set of prime numbers between 3 and 23 including the extremities. What is the probability the pair adds to 24?

6.13. An urn contains 5 balls marked with numbers from 2, 3, 4, 5 or 6. Three balls are drawn at random, one at a time. Now a three digit number is constructed with the digits engraved on the drawn balls in the sequence. What is the probability that the number so constructed is a multiple of 4?

6.14. A dart board has the configuration with points as shown in the figure below. If two darts are thrown at random, what is the probability of getting a total score of 4 points? Note that the given triangle is an equilateral triangle.

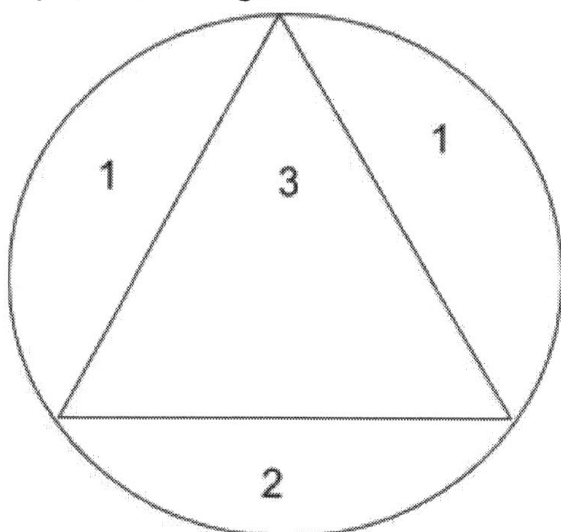

6.15. There are 5 cards with denominations 2, 4, 6, 7 and 9 . If one card is picked at random, put back, and second one is picked at random. What is the probability of getting a total more than 13?

6.16. A dart board has an two equilateral triangles drawn on it as shown in the given figure. The centroids of two triangles match and the scaling factor of the smaller vs larger triangle is x. Three circum-radii divide each triangle into the three congruent lobes with point values shown. The probability that a dart will hit a given region is proportional to the area of the region. When two darts hit this triangular area, the score is the sum of the point values in the regions. What is probability that the score is odd in terms of x?

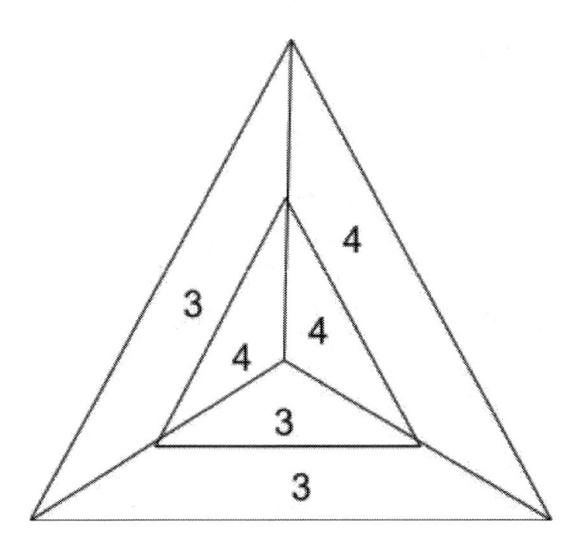
4
3
4
4
4
3
3

P7. Smallest and Largest Numbers

1. What is the difference between the smallest and largest 4 digit numbers that are divisible by 10, 15 and 23?
2. What is largest a four digit number that is a square and has no prime factors that are less than 60?
3. What is the largest 3 digit number that leaves a residue of 24 when divided by 93 and is a multiple of 5?
4. LCM and HCF of two numbers are 58 and 870 respectively, and the larger number is not divisible by the smaller number. What are the two numbers?
5. What is the largest number that leaves the same remainder in each case when 78, 225 and 414 are divided by the number?
6. What is largest natural number that divides the product of three consecutive integers exactly?
7. What is largest natural number that divides the product of 4 consecutive natural numbers exactly?
8. What is the minimum number of soldiers that must join a group of 9778 soldiers so that they can form a square array?
9. 400 scouts were taking part in a parade. In the middle of the drill, they proceed to form a triangular pattern(as shown in the figure) , where k-th row has k number of

scouts. Minimum how many scouts should stay back so that all the rows are completely filled?

10. Find the greatest number of three digits which when divided by 10, 16 and 24 leaves the remainders 7, 13 and 21 respectively.

Hints and Answers
to
Selected Problems

Hints

Exponents

1.1. x= - 3/2 , y = -2 [Hints: use prime factorization of both sides of the equation]

1.2. No solutions

1.3. x=2, y = 4

1.4. a

1.5. Apply basic law of exponents and get the equation with smallest factor.

1.6. -1

1.7. $\frac{-16}{3}$

1.8. 200 [Hints: $1 + 2 + 3 + \ldots + n = \frac{n(n+1)}{2}$]

1.9. 40.

1.10. $x = \frac{7}{12}$ $and\, y = \frac{5}{16}$ [Hints: Solve two simultaneous equations formed after equating powers of 2 and 3].

1.11. 2 [Hints: unit digit of 3^{4k} is 1, unit digit of 3^{4k+1} is 3, unit digit of 3^{4k+2} is 9, and so on]

Using Standard Formulas of Squares and Sums

[Hints: apply
$a^3 - b^3 = (a-b)(a^2+ab+b^2)$, $a^3 + b^3 = (a+b)(a^2-ab+b^2)$, $a^2 - b^2 = (a-b)(a+b)$, $(a+b)^2 = a^2+2ab+b^2$]

2.1. 98
2.2. 83
2.3. 76
2.4. 13
2.5. 52
2.6. 4.0
2.7. 700
2.8. 100
2.9. 16,17
2.10. -1.5
2.11. 24

Geometry

3.1. 4cm [Hints: Area of an equilateral triangle with side $= a\ cm$, is $\frac{\sqrt{3}a^2}{4} cm^2$]

3.2. 150 [Hints: How many triangles are formed If diagonals are drawn starting from the same vertex?]

3.3. $\frac{1}{4}b\sqrt{4l^2 - b^2}$

3.4. 2.5

3.5. 25%

3.6. ⅓ . ⅓ , 1/18 , 5/18

3.7. 21 square units.

3.8. 15cm [Hints: Prove Δ BAP ~ Δ BQC]

3.9. 32 cm^2. TODO

3.10. $\frac{144}{13}$ [Hints ΔABC ~ ΔBPC by AA rule of similar triangles ⇨ $\frac{AC}{BC} = \frac{AB}{BP} = \frac{BC}{PC}$]

3.11. 48° , 72°, 96°, 66°, 48° [Hints Angle subtended by each arc at the center = 360/ 15 = 24°]

3.12. 2cm [Hints: Use the Proof of Pythagorean theorem].

3.13. $\frac{1}{2}$ [Hints: Use definition of incenter, circumcenter, Ratio of the circumference of two circles = ratio of the radii of two circles.]

3.14. $(\sqrt{3} - \frac{\pi}{3})$ 16 sq units [Hints: Area of shaded region = Area of the circle - Area of inner triangle. }

3.15. 15 cm. ∠PAN = 90 - ∠APN

3.16. 5π *sq units* [Hints: Find the radius. The Origin O(0,0), point of tangency P and the centre C(5,0) form a right triangle with right angle at P.]

Functions

4.1. $f(x) = -\frac{20}{49}x^2 + \frac{260}{49}x - \frac{600}{49}$ (x-3) and (x-10) are factors in function.

4.2. $f(x) = \frac{1}{8}x^3 - \frac{3}{8}x^2 + \frac{5}{4}x$

4.3. $f(2y+1) = 8y^3 + 12y^2 + 20y + 3$

4.4. $f(x) - 6x - 9$ [Hints: Find the value of x in terms of t, such that $\frac{3+x}{2} = t$]

4.5. $f(x) = \frac{25}{9}x^2 - \frac{55}{9}x + \frac{2}{9}$

4.6. domain { x | x is a real number}, and range $\{y|\ y \text{ is a non} - \text{negative real number}\}$.

4.7.

D:all real numbers R:positive numbers	D: all real number except 3 R: all positive real numbers	D:all real numbers except 4 R:all negative real numbers
D: All real numbers R: All real numbers ≥4	D: all real numbers except 0 R: all real numbers except 0	D: All real numbers R: all real numbers ≥6
D: all real numbers ≥ 3 R:all real numbers ≥ 0	D: all real numbers R: all real numbers	D:All real numbers ≥ 0 R: all real numbers ≥ 0

4.8. y intercept is 13 and slope = -3.

4.9. $x = -\frac{5}{3}$

4.10. $f(n) = \frac{n(n+1)}{2}$ Factorize the given equation.

4.11. $a_1 = 5(1) - 11 = -6$, d = 5

4.12. $a_1 = 25$ and d =7 [Hints: use the standard form for a_n in an AP]

4.13. 7

4.14. (2, -4)

4.15. 7, 10 and 12

4.16. s=54, t= -1 [[Hints: An exponential function is of the form $f(x) = ab^x$ where a and b are real numbers.]

4.17. $a_n = -5 + 1.5(n-1)$

4.18. $3y - 4x - 46 = 0$ [Hints: Given line in slope intercept form : y = -(¾)x + 20/4.]

4.19. $f(n) = f(n-1) - \frac{7}{20}$ $and\, f(1) = \frac{3}{5}$ [Hints: Find a1 and a(n) - a(n-1)]

4.20. $k = 13$

4.21. [Hints: Write S_n in reverse order, and add up with original form of S_n]

4.22. 54th

4.23. 6

4.24. 67

4.25. 3

Speed, Time, Work

5.1. $1.2\ sec$ [Hints: time taken to cross a pole = time taken traverse its own length since pole width is negligible, as compared to train.]

5.2. 1.5min

5.3. Bobby 4 rounds and Anne 5 rounds.

5.4. 140m

5.5. 24 days

5.6. A in 14.4 days, B in $10\frac{2}{7}\ days$ and C in 72 days.

5.7. 15 hrs [Hints: Net filling rate is: filling rate - emptying rate]

5.8. $5\frac{4}{6}min$.

5.9. 20 kmph
5.10. 1.2 km.
5.11. 24 kmph [Hints: Assume the length of the trail as something]
5.12. 12 sec
5.13. 150 sec
5.14. 13 m/s [Hints: For the same distance covered, Speed is inversely proportional to time]
5.15. 14.4 s
5.16. 2 Hrs.
5.17. $288\ kmph$
5.18. $16\ kmph$ [Hints: Calculate ratio of speed and ratio of distance to be covered.]
5.19. 52 Hrs [Hints :Use unitary method- find out time required for 1 work to be finished by 1 scheduled time in 8 hours].
5.20. 1 day.

Probability

6.1. $0.8(1 - \frac{1}{\pi})$ [Hints: Probability of a dart that lands on a dart hitting a given area = $\frac{Area\ of\ the\ shaded\ area}{Are\ of\ the\ dart\ board}$]
6.2. $\frac{7}{24}$
6.3. $\frac{127}{610}$.
6.4. 0.1536 [Hints : 6 possible permutation of wins and losses.].
6.5. 18

6.6. $\frac{2}{7}$ [Hints: Find number of permutations with one boy on both sides,]

6.7. $\frac{1}{5}$

6.8. $\frac{6}{35\times37}$

6.9. $\frac{5}{36}$

6.10. $\frac{2}{7}$ [Hints: There are 6 sequences possible. Find probability of each sequence]

6.11. $\frac{5}{12}$ [Hints: Probability of arriving within a time slot =(length of time slot)/(total available time)]

6.12. $\frac{3}{28}$

6.13. $\frac{3}{10}$

6.14. $(1 - \frac{3\sqrt{3}}{4\pi})\ (\frac{1}{3} + \frac{3\sqrt{3}}{4\pi})$ Get probability of dart in each area and then probability of 4.

6.15. $\frac{1}{5}$.

6.16. $\frac{2}{9}(2 + x^2 - x^4)$

Smallest and Largest Numbers

7.1. 8280 [Any number divisible by given set of numbers is LCM of those numbers. 9409]

7.2. 9409

7.3. 675

7.4. 174 and 290 [Hints: What are factors for LCM * HCF]

7.5. 21 [Hints :Since remainder is same, difference of numbers is divisible by same number.]

7.6. 6 [Hints: If there are three consecutive numbers one of those has to be divisible by three and one has to be even number.]

7.7. 24 [Hints: If there are four consecutive numbers, one of those have to be divisible by three, at least 2 even numbers should be there.]

7.8. 222

7.9. 6

7.10. 237 [Note the divisor -remainder difference, each is 3.]

Solutions

S1.Exponents

1.1. We can see both sides of the equation have 2 and 5 are smallest prime factor. This gives you a hint that you need to express both sides in terms of 2 and 5.

$LHS = 5 \cdot 50^{y} = 5 \cdot (2 \cdot 25)^{y} = 5 \cdot (2 \cdot 5^{2})^{y} = 2^{y} \cdot 5^{2y+1}$

$RHS = 2 \cdot 4^{x} \cdot 5^{2x} = 2 \cdot 2^{2x} \cdot 5^{2x} = 2^{2x+1} \cdot 5^{2x}$

As LHS = RHS, and have the same bases.

Equating powers of 2 you get

$$y = 2x + 1 \quad .....(i)$$

Equating powers of 5, you get

$$2y + 1 = 2x \;....\; (ii)$$

$\Rightarrow y - 1 = 2x$ [*From* (*i*)] *and* $2x = 2y + 1$ [*Equation* (*ii*)]

$\Rightarrow y - 1 = 2y + 1$

$\Rightarrow y = -2,\ x = (y - 1)/2 = -3/2$

1.2. Given: $10^{y} = 8 \cdot 4^{x} \cdot 5^{2x}$

or $(2 \cdot 5)^{y} = 2^{3} \cdot 2^{2x} \cdot 5^{2x} = 2^{2x+3} \cdot 5^{2x}$

$2^{y} \cdot 5^{y} = 2^{2x+3} \cdot 5^{2x}$

Equating exponents from the both sides of the equality, we get

$y = 2x + 3$ *and* $y = 2x$

These two equations have no solutions,

1.3. Given $\sqrt[x]{10^{y} \cdot 25} = 500$

$$or\ \sqrt[x]{2^y \cdot 5^y \cdot 5^2} = 4 \times 125$$

$$or\ \sqrt[x]{2^y \cdot 5^{y+2}} = 2^2 \cdot 5^3$$

$$or\left(\sqrt[x]{2^y \cdot 5^{y+2}}\right)^x = \left(2^2 \cdot 5^3\right)^x$$

$$or\ 2^y \cdot 5^{y+2} = 2^{2x} \cdot 5^{3x}$$

Equating exponents from the both sides of the equality, we get $y = 2x\ and\ y + 2 = 3x$

Or $2 = x\ ,\ and\ y = 2 \times 2 = 4$

Or solution is x=2 and y = 4.

1.4. In the prime factorization of a perfect cube's, a prime factor repeats a multiple of 3 number of times. So if the prime factorization is expressed as a product of powers of prime factors, each power should be a multiple of 3.

a. $4^{13} = 2^{26}$, 26 is not divisible by 3
b. $8^{13} = 2^{3\times13}$, the power is a multiple of 3
c. $20^{27} = (20^9)^3$, cube
d. $4^{100} \times 2^{100} = 2^{300}$, 300 is a multiple of 3

(a) Is not a perfect cube.

1.5. This question is asking you to apply basic laws of exponents

$$\sqrt[5]{16^{2n}}\sqrt[5]{64^{2n}} = \sqrt[5]{(2^4)^{2n}}\sqrt[5]{(2^6)^{2n}} = \sqrt[5]{2^{8n} \cdot 2^{12n}} = \sqrt[5]{2^{20n}} = 2^{4n} = 16^n$$

1.6. LHS=

$$\sqrt[3]{x^5} \times \sqrt[3]{x^4} \times x^k = x^{\frac{5}{3}} \cdot x^{\frac{4}{3}} \cdot x^k = x^{\frac{5}{3}+\frac{4}{3}+k} = x^{\frac{5+4}{3}+k} = x^{3+k}$$

Given, this equals x^2.

$\therefore 2 = 3 + k \; or \; k = -1$

1.7. Given $4^{2/x} = \frac{1}{\sqrt[4]{8}}$

$Or\ (2^2)^{2/x} = 8^{-1/4}$

$Or\ 2^{4/x} = (2^3)^{-1/4} = 2^{-3/4}$

$Or\ \frac{4}{x} = \frac{-3}{4}$

$Or\ 4 \times 4 = -3x \Rightarrow x = \frac{-16}{3}$

1.8. LHS = $4 \cdot 4^2 \cdot 4^3 \cdot 4^4 \cdot \ldots \cdot 4^{24} = 4^{1+2+..+24}$

$= 4^{24(24+1)/2}$ [*Using* $1 + 2 + 3 + \ldots + n = \frac{n(n+1)}{2}$]

$= 4^{12\times 25} = 4^{300} = 2^{600}$

RHS = $8^x = 2^{3x}$

Using LHS=RHS, $2^{600} = 2^{3x}$ you get 600 = 3x or x =600/3 = 200.

1.9. Given $\sqrt[x]{q^y} \times \sqrt[y]{q^x} = q^{\frac{10}{xy}}$

$$Or\ q^{\frac{y}{x}} \times q^{\frac{x}{y}} = q^{\frac{y}{x}+\frac{x}{y}} = q^{\frac{10}{xy}}$$

$$Or\ \frac{y}{x} + \frac{x}{y} = \frac{10}{xy}$$

$$Or\ \frac{y^2+x^2}{xy} = \frac{10}{xy}$$

$$Or\ y^2 + x^2 = 10$$

$\therefore 4(x^2 + y^2) = 40$

1.10. Given $\sqrt{8 \cdot 3} \cdot \sqrt[4]{27 \cdot 2} = 8^x \cdot 81^y$

$$Or\ \sqrt{2^3 \cdot 3} \cdot \sqrt[4]{3^3 \cdot 2} = \left(2^3\right)^x \left(3^4\right)^y$$

$$Or\ 2^{\frac{3}{2}} \cdot 3^{\frac{1}{2}} \cdot 3^{\frac{3}{4}} \cdot 2^{\frac{1}{4}} = \left(2^3\right)^x \left(3^4\right)^y$$

$$Or\ 2^{\frac{3}{2}+\frac{1}{4}} \cdot 3^{\frac{1}{2}+\frac{3}{4}} = \left(2^3\right)^x \left(3^4\right)^y$$

$$Or\ 2^{\frac{7}{4}} \cdot 3^{\frac{5}{4}} = 2^{3x}\ 3^{4y}$$

$Or\ \frac{7}{4} = 3x\ and\ \frac{5}{4} = 4y$, equating powers of the same bases

$\therefore\ x = \frac{7}{12}\ and\ y = \frac{5}{16}$

1.11. Let's look at first few powers of 3, starting from 1 are : 3, 9, 27, 81, 243,.. Unit digits are in the sequence 3,9,7,1,3,9.. [repeating after every 4].

I.e. If the power is 4k, unit digit is 1; power is 4k+1, unit digit is 3; power is 4k+2, unit digit is 9; and so on...

Similarly, $4^1 = 4,\ 4^2 = 16,\ 4^3 = 64,\ 4^4 = 256$.. Unit digits of powers of 4 starting with power 1 are 4, 6, 4,... i.e. every odd power of 4 has unit digit 4, and even power 6.

Now $1042 = 4 \times 260 + 2$, So unit digit of 3^{1042} is 9.

Similarly unit digit of $4^{99}(odd\ power)$ would be that of 4

$\therefore$ Unit digit of $3^{1042} \cdot 4^{99}$ is unit digit of 4×3 i.e. 2.

S2.Using Standard Formulas of Squares and Sums

2.1. Given $p + 1/p = 10$

Squaring both sides we get $(p + 1/p)^2 = 10^2$

$p^2 + (1/p)^2\ 2p(1/p) = 100$

$p^2 + 1/p^2 + 2 = 100$

$p^2 + 1/p^2$ = 100-2 = 98. (Ans)

2.2. $$p^2 + \frac{1}{p^2} = p^2 + \frac{1}{p^2} - 2p(\tfrac{1}{p}) + 2p(\tfrac{1}{p})$$

$$= (p - \tfrac{1}{p})^2 + 2 = 9^2 + 2 = 83$$

2.3. Use the standard formula

$$a^3 - b^3 = (a - b)(a^2 + ab + b^2)$$

$$p - 1/p = 4$$

$$\Rightarrow p^2 + p(1/p) + \frac{1}{p^2}$$

$$= p^2 - 2p(1/p) + \frac{1}{p^2} + 3p(1/p)$$

$$= (p - 1/p)^2 + 3 = 4^2 + 3 = 19$$

$$\therefore p^3 - \frac{1}{p^3} = (p - 1/p)(p^2 + p \cdot \tfrac{1}{p} + \tfrac{1}{p^2})$$

$$= 4 \cdot 19 = 76$$

2.4. Use the standard formula

$$a^3 + b^3 = (a+b)(a^2 - ab + b^2)$$

$$p + \frac{1}{p} = 4 \Rightarrow p^2 - p \cdot \frac{1}{p} + \frac{1}{p^2}$$

$$= p^2 + 2p(1/p) + \frac{1}{p^2} - 3p(1/p)$$

$$= (p + 1/p)^2 - 3 = 4^2 - 3 = 13$$

$\therefore$

$$p^3 + \frac{1}{p^3} = (p + \frac{1}{p})(p^2 - p \cdot \frac{1}{p} + \frac{1}{p^2}) = 4 \cdot 13 = 52$$

2.5. $$\frac{6.3\times6.3\times6.3 - 2.3\times2.3\times2.3}{6.3\times6.3 + 6.3\times2.3 + 2.3\times2.3}$$

$$= \frac{6.3^3 - 2.3^3}{6.3^2 + 6.3\times2.3 + 2.3^2}$$

$$= \frac{(6.3 - 2.3)(6.3^2 + 6.3\times2.3 + 2.3^2)}{6.3^2 + 6.3\times2.3 + 2.3^2}$$ [using $a^3 - b^3 = (a-b)(a^2 + ab + b^2)$]

= (6.3 - 2.3) = 4.0 [Ans]

2.6. You can see that the numerator is of the form a^3+b^3, so you get the hint that you need to use the formula for a^3+b^3. Also note the denominator is of the form of a factor of a^3+b^3.

$$\frac{499\times499\times499 + 201\times201\times201}{499\times499 - 499\times201 + 201\times201} = \frac{(499 + 201)(499\times499 - 499\times201 + 201\times201)}{(499\times499 - 499\times201 + 201\times201)}$$

$$= (499 + 201) = 700$$

2.7. Let $a \text{ and } b$ be the two numbers, with being the larger one.

Given $a^2 + b^2 = 125 \text{ and } ab = 37.5$

$(a+b)^2 = a^2 + b^2 + 2ab = 125 + 2 \times 37.5 = 200$

$(a+b) = \sqrt{200} = 10\sqrt{2}$

Similarly,

$(a-b)^2 = a^2 + b^2 - 2ab = 125 - 2 \times 37.5 = 50$

$(a-b) = \sqrt{50} = 5\sqrt{2}$

∴ difference of the squares of these two numbers =

$a^2 - b^2$

$= (a-b)(a+b) = 10\sqrt{2} \times 5\sqrt{2} = 100$

2.8. Let the smaller number be n. So the larger number is n+1

Difference between their squares =

$(n+1)^2 - n^2 = (n+1+n)(n+1-1) = 2n+1$

Given 2n+1 = 33

$Or\ 2n = 33 - 1 = 32 \Rightarrow n = 32/2 = 16$

∴ The numbers are 16 and 17.

2.9. Let the numbers be $a \text{ and } b$.

Given $a^2 + b^2 = 200 \quad \text{and } a^2 - b^2 = 40$

$2a^2 = 200 + 40 = 240 \Rightarrow a^2 = 240 \div 2 = 120$

$b^2 = 200 - a^2 = 200 - 120 = 80$

$Or\ a^2b^2 = 80 \times 120 = 9600$

$Product\ of\ two\ numbers = ab = \pm\sqrt{9600} = \pm 40\sqrt{6}$

2.10. $LHS = 2x^2 - 2xy + y^2 + 3x + 2.25$

$= x^2 - 2xy + y^2 + x^2 + 3x + 2.25$

$= (x - y)^2 + (x + 1.5)^2$

Square of a real quantity is always greater or equal to 0.

For RHS to be 0, each square term has to be 0.

=> $(x - y) = 0\ and\ (x + 1.5) = 0$

=> $x = y\ and\ x = -1.5$

=> $x = -1.5,\ y = -1.5$

2.11. Given $a^2 + b^2 + c^2 - ab - bc - ca = 0$

Or $2(a^2 + b^2 + c^2 - ab - bc - ca) = 2(0) = 0$

Or $2a^2 + 2b^2 + 2c^2 - 2ab - 2bc - 2ca = 0$

Rearranging LHS we get

$a^2 - 2ab + b^2 + b^2 - 2bc + c^2 + c^2 - 2ca + a^2 = 0$

$(a - b)^2 + (b - c)^2 + (c - a)^2 = 0$

As squares of real numbers cannot be negative, to each square has to be 0 to hold the equality above

I.e. $(a - b)^2 = 0,\ (b - c)^2 = 0\ and\ (c - a)^2 = 0$

$\Rightarrow (a - b) = 0,\ (b - c) = 0\ and\ (c - a) = 0$

$\Rightarrow a = b = c$

$\therefore\ a + 2b + 3c = a + 2a + 3a = 6a = 6 \times 4 = 24$

S3. Geometry

3.1. Area of an equilateral triangle with side a cm is $\frac{\sqrt{3}a^2}{4}cm^2$

$\frac{\sqrt{3}a^2}{4} = \frac{\sqrt{3}}{4}16 \Rightarrow a^2 = 16cm^2 \Rightarrow a = 4$

∴ length of a side = 4cm

3.2. Sum of all the interior angles of an n-sided polygon:
If diagonals are drawn starting from the same vertex, n-2 triangles are formed
Sum of all the internal angles of the polygon = (sum of angles of n-2 triangle above) = (n-2)**π** rad = (n-2)180 degree.
The measure of each angle of a regular dodecagon = (12-2)180/12 = 150 degree.

3.3. Try to find the height of the isosceles triangle in terms of l and b.

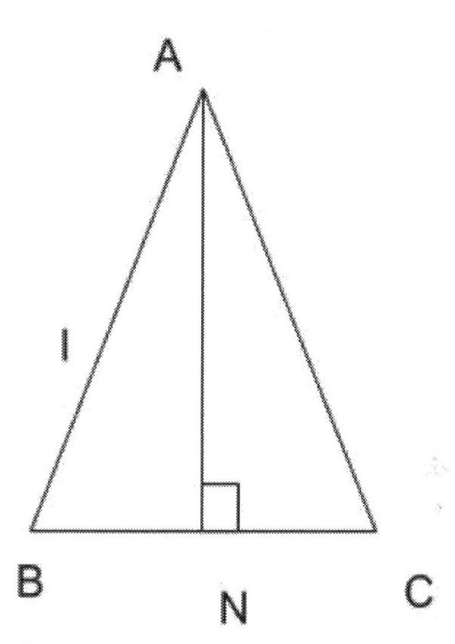

ΔABC, in the figure, is an isosceles triangle with AB and AC being the equal sides.

Let AN⊥BC at N. Between the two right triangles ΔABN and ΔACN,

AB = AC (by definition of the given triangle) and are the hypotenuses, AN =AN.

By RHS axiom ΔABN ≅ΔACN ∴ BN = CN = b/2

In right triangle ΔABN , AN =

$$\sqrt{l^2 - (b/2)^2} = \frac{1}{2}\sqrt{4l^2 - b^2}$$

∴ Area of ΔABC =

$$\frac{1}{2}\ bases \times heigh = \frac{1}{2}b\frac{1}{2}\sqrt{4l^2 - b^2} = \frac{1}{4}b\sqrt{4l^2 - b^2}$$

(Ans)

3.4. Given: Square with ABCD is inscribed in a bigger square PQRS . Vertex A lies on the side PQ such that PA=a cm and QA = b cm.

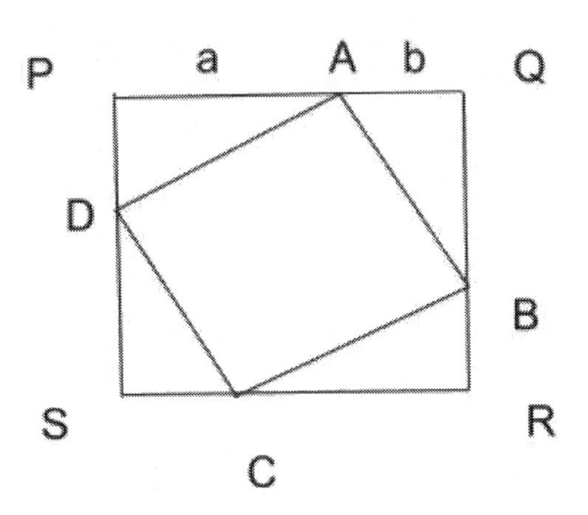

Between **Δ**APD and **Δ**BQA,

$\measuredangle$DAP = 180° - 90° - $\measuredangle$QAB = 90° - $\measuredangle$QAB = $\measuredangle$ABQ

$\measuredangle$P = 90° = $\measuredangle$Q

AD = AB [sides of the same square]

ΔAPD ≅ **Δ**BQA , by AAS axiom

∴ QB = a cm (side corresponding to PA)

Using the same same logic, we can show that all the triangles formed between two squares are congruent.

∴ Area between 2 squares = 4 (area of **Δ**BQA) = 4 (ab/2) = 2ab cm^2

Also area between 2 squares = 18 - 13 = 5 cm^2

∴ ab = 5/2 = 2.5 (Ans)

3.5. Given: Two diagonals of the rectangle ABCD intersect at E. A straight line through E intersects AC and BD at P and Q respectively.. Hence find the percent area of the rectangle covered by the shaded region.

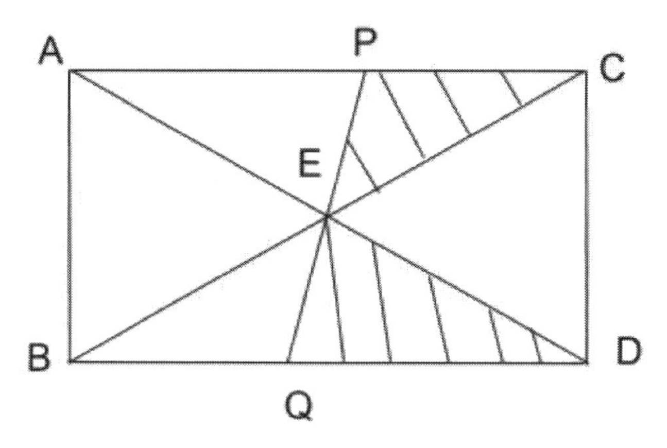

Between: ΔAEP and ΔDEQ

AE = ED

∠AEP = ∠DEQ (Vertically opposite angles)

∠EPA = ∠ EQD (Internal alternate angles for PQ intersecting AC||BD)

By AAS axiom , $\Delta AEP \cong \Delta DEQ$

Area of the shaded area

= area(**Δ** DEQ) + Area(**Δ**EPC)

= area (**Δ**AEP) + Area(**Δ**EPC)

= Area (**Δ**AEC)

= 0.25 Area(rectangle ABCD)

∴ Percent area = 25% (Ans).

3.6. Given: In rectangle ABCD a line intersects sides AB and BC at E and F respectively. Also given is that AE : BE = 2: 1 and CF : BF = 2:1. Compute areas of $\Delta AED,\ \Delta DFC\ and\ \Delta BEF$ as fraction of area of the rectangle. Now find the area of the shaded region as fraction of the area of the rectangle.

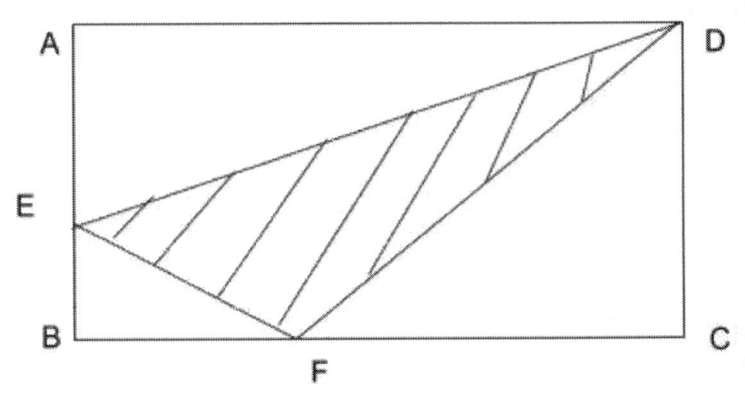

Let AD = l and AB = b

AE =⅔ b , BE = ⅓ b, CF = ⅔ l, BF = ⅓ l

Area(ΔAED) = ½ AE × AD = ½ (⅔ b) l = ⅓ bl = ⅓ (area of rectangle) [Ans]

Area(ΔBEF) = ½ CF × CD = ½ (⅔ l) b = ⅓ bl = ⅓ (area of rectangle) [Ans]

ARea(ΔBEF) = ½ BE × BF = ½ (⅓ b) (⅓ l) = $\frac{1}{18}$ bl = $\frac{1}{18}$ (area of rectangle) [Ans]

Area (shaded triangle) = area of rectangle - area(unshaded region)

= bl - (⅓ bl + ⅓ bl + $\frac{1}{18}$ bl) = bl - $\frac{13}{18}$ bl = $\frac{5}{18}$ bl

$= \frac{5}{18}$ (Area of rectangle) [Ans]

3.7. Given: Areas of the triangles are specified in the figure (what is the question?)

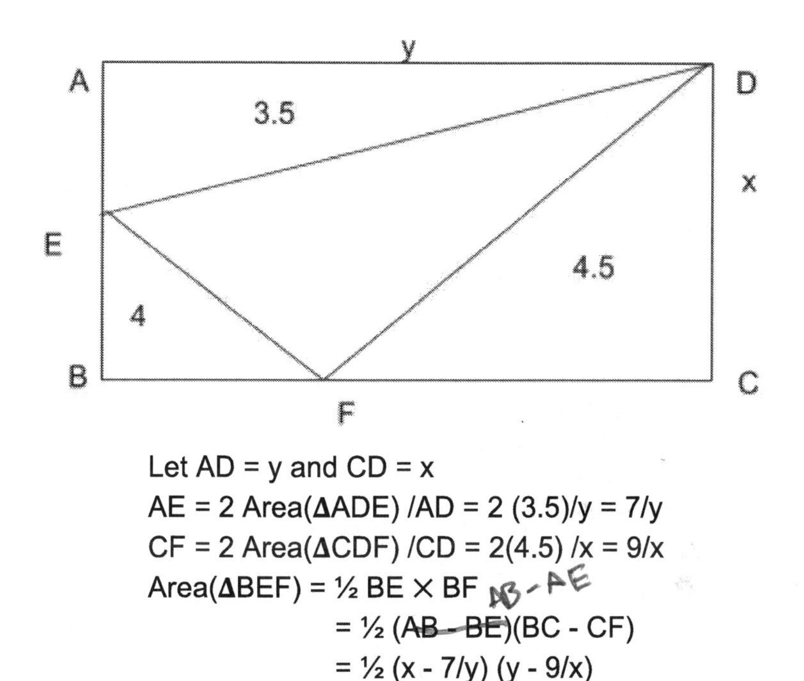

Let AD = y and CD = x

AE = 2 Area(ΔADE) /AD = 2 (3.5)/y = 7/y

CF = 2 Area(ΔCDF) /CD = 2(4.5) /x = 9/x

Area(ΔBEF) = ½ BE × BF

= ½ (AB - BE)(BC - CF)

= ½ (x - 7/y) (y - 9/x)

= ½ (xy -7 -9 + 63/cy)

= ½ (xy -16 + 63/xy)

Per figure, Area(ΔBEF) = 4

½ (xy -16 + 63/xy) = 4

Or (xy + 63/xy) = 18

Or $(xy)^2$ -24(xy) + 63 = 0

Or (xy -21)(xy -3) = 0

Or xy = 21 , 3

Area of the rectangle > 12, so discard the root 3.

∴ Area of the rectangle 21 sq units. [Ans]

3.8. Given: AQ = AC, AP is the bisector of ∠BAC .

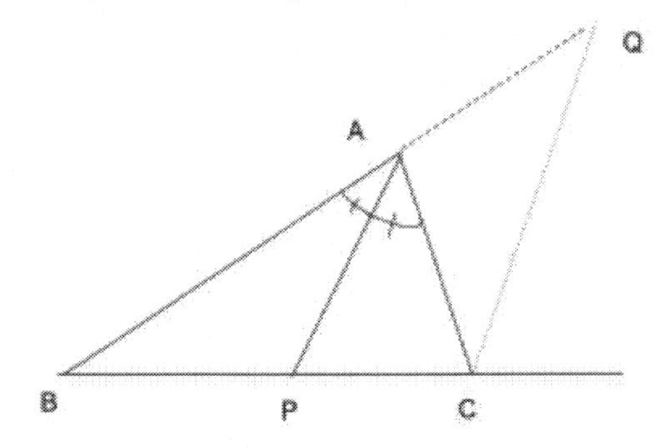

Between Δ BAP and Δ BQC

∠B = ∠B

∠BAP = ½ ∠BAC

= ½ (∠ACQ +∠AQC) [exterior angle of triangle = sum of other two angles]

= ½ (2 ∠AQC) [∵ Angles opposite to equal sides AC ,AQ in Δ ACQ]

= ∠AQC

∴ By AA rule of similar triangles, Δ BAP ~ Δ BQC

Applying corresponding side ratio,

CQ = $\frac{BQ}{BA} \times AP = \frac{(BA+AQ)}{BA} \times AP = \frac{(BA+AC)}{BA} \times AP$

$= \left(1+\frac{AC}{AB}\right) AP = (1+\frac{1}{2})\ 10\ cm = 15\ cm$ [Ans]

3.9. Given: ABCD is a square and ΔAPD is a triangle right angled at P. AP=6cm and PD= 8cm.

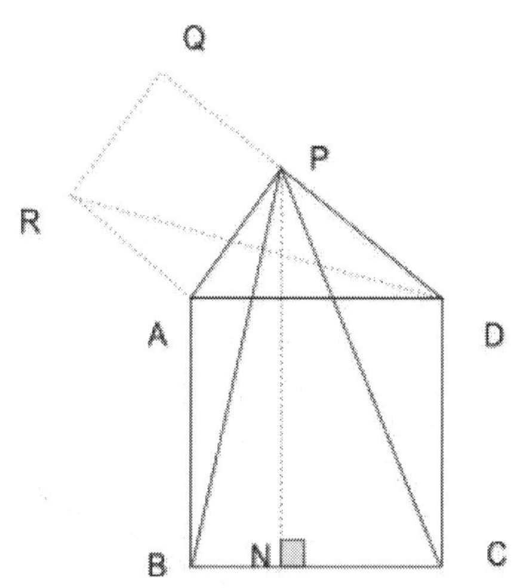

This problem is based on Euclid's proof of Pythagorean Theorem. You will need to do constructions as done in the proof.
With AP as a side let's complete the square APQR.
Also let's drop PN⊥ BC.

Between Δ ADR and Δ ABP,
AD = AB .... Sides of square ABCD
AR = AP .... Sides of square APQR
∠DAR = ∠PAR + ∠DAP = ∠DAB + ∠DAP = ∠BAP
By SAS axiom, Δ ADR ≅ Δ ABP
Area (Δ ABP) = Area (Δ ADR)
= ½ Area(square APQR) [as they have same base and height
= ½ $(\overline{AB})^2$ =18 cm^2 [Ans]

Area (Δ CDP) = ½ Area(square on PD) = ½ $(8)^2$ = 32 cm^2 [Ans]

3.10. Given: In triangle ΔABC, ∠B = 90°. BP is perpendicular to AC at point P. Length of BC = 5 cm and length of BA = 12cm.

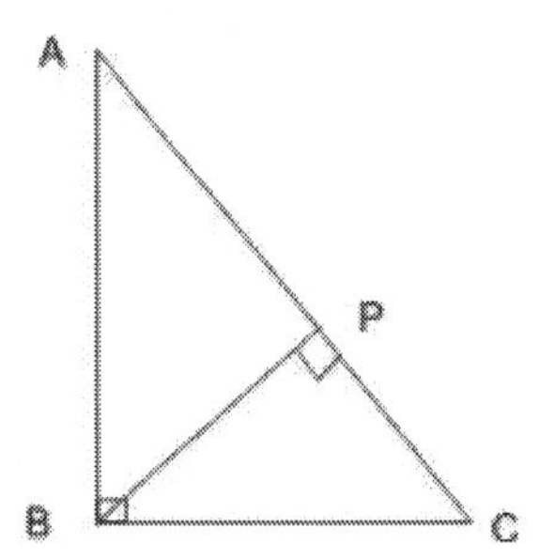

To Find: BP, PC and AP.

Between ΔABC and ΔBPC

∠B = ∠P = 90°

∠C = ∠C

ΔABC ~ ΔBPC by AA rule of similar triangles

$$\frac{AC}{BC} = \frac{AB}{BP} = \frac{BC}{PC}$$

$$\overline{AC} = \sqrt{5^2 + 12^2} = 13$$

$$\overline{BP} = \frac{\overline{AB}}{AC} \times \overline{BC} = \frac{12}{13} \times 5 = \frac{60}{13}$$

$$\overline{PC} = \frac{\overline{BC}}{AC} \times \overline{BC} = \frac{5}{13} \times 5 = \frac{25}{13}$$

Using similarity between ΔABC ~ ΔAPC

$$\overline{AP} = \frac{\overline{AB}}{AC} \times \overline{AB} = \frac{12}{13} \times 12 = \frac{144}{13}$$

3.11. Given:A circle with centre O is divided into 15 equal arcs and marked A through O in the clockwise order.
Angle subtended by each arc at the center = 360/ 15 = 24°
∠AOC = angle subtended by two consecutive arcs = 48°
∠DOG = angle subtended by three consecutive arcs DE, EF, FG = 72°
∠HOL = angle subtended by 4 consecutive arcs = $4 \times 24 = 96^o$
In ΔAOC, ∠ACO = ∠CAO.
So ∠ACO = ½ (180° - ∠AOC) = ½ (132) = 66° [Ans]
Similarly ∠HLO = ½ (180° - ∠HOL) = ½ (180° - 96°) = 48° [Ans]

3.12. This problem is based on the proof of the Pythagorean theorem given by Ancient Indian mathematicians. If we rearrange four copies of the given triangle, we get the following figure:

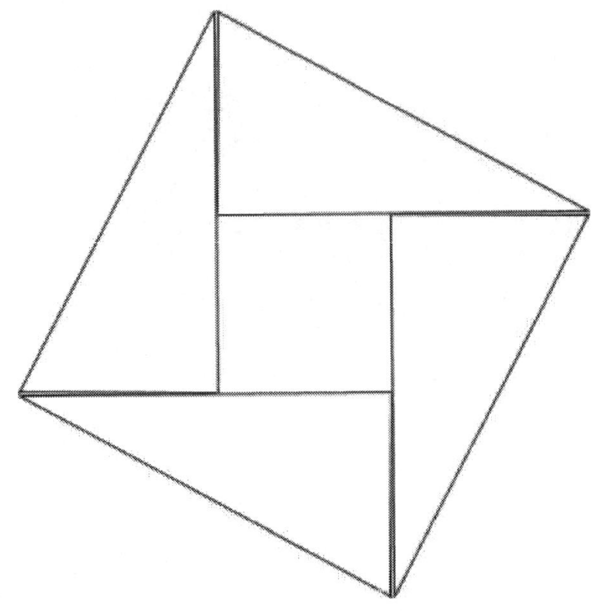

Clearly inner square's area is square of difference of two sides of the triangle.

Area of the outer square = $(\sqrt{104})^2 = 104\ cm^2$

Total area of four triangles = $10^2 = 100cm^2$

Area of the inner square = $104 - 100 = 4\ cm^2$

$\therefore$Difference of length of two sides = $\sqrt{4} = 2$ cm

3.13. Given: A circle is inscribed in an equilateral triangle, and a larger circle circumscribes the triangle

Draw the figure:

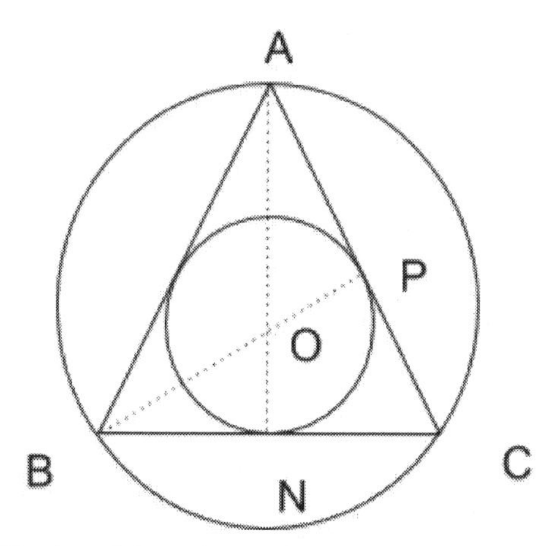

Ratio of the circumference of two circles = ratio of the radius of two circles.

Let two angle bisectors BP and AN meet at O.

For an equilateral triangle, BN = NC, and ∠ANB = 90°

Between ΔAPO and ΔANB,

∠P = ∠N = 90°

∠PAO = ∠NAB = 30°

ΔAPO ~ ΔANB

$\frac{OP}{AO} = \frac{BN}{AB} = \frac{1}{2}$

Inner circumferences : Outer circumference = $\frac{1}{2}$ Ans

3.14. Given:An equilateral triangle is inscribed in a circle, and the circle is inscribed in a bigger equilateral triangle. The area of the region between the smaller triangle and the circle is $(\frac{\pi}{3} - \frac{\sqrt{3}}{4})\,16$ square units.

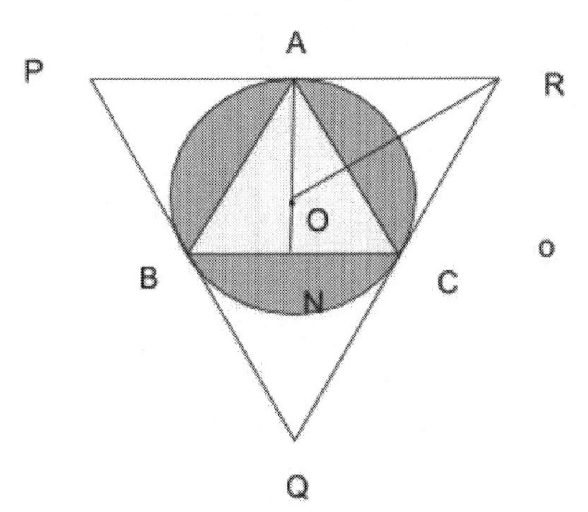

$AO = \frac{2}{3} AN = \frac{2}{3} \times \frac{\sqrt{3}}{2} AB = \frac{AB}{\sqrt{3}}$

Let length of AB be a.

Area of the circle = $\frac{\pi}{3} a^2$

Area of the inner triangle = $\frac{\sqrt{3}}{4} a^2$

Area of shaded region = $(\frac{\pi}{3} - \frac{\sqrt{3}}{4})\, a^2$

Shaded area is $(\frac{\pi}{3} - \frac{\sqrt{3}}{4})\, 16$

$\therefore a^2 = 16$ or $a = 4$

PQ = 2 PB = 2 AB = 2a

Area of the larger triangle = $\frac{\sqrt{3}}{4} (2a)^2 = \sqrt{3}\, a^2$

Area between the larger triangle and the circle = $\sqrt{3}\,a^2 - \frac{\pi}{3}a^2 = (\sqrt{3} - \frac{\pi}{3})\,16$ sq units

3.15. Given: From a point P on the circle $\odot O$ with radius 10 cm is drawn a line segment $\overline{PN}$ perpendicular to the diameter AB at N. ON is 5cm

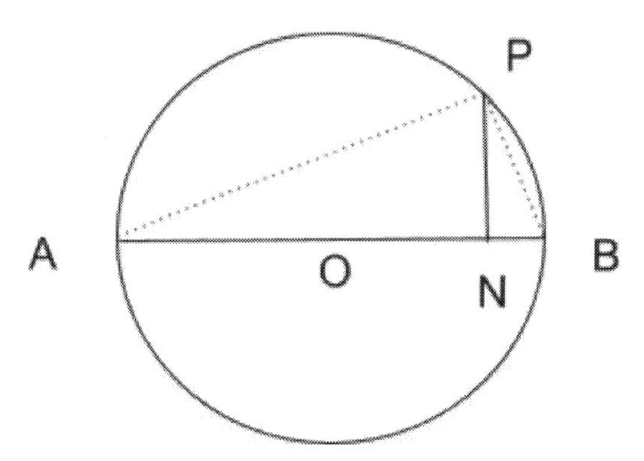

Between ΔAPN and ΔPBN

∠PAN = 90 - ∠APN [∵ ∠PNA = 90° Given, and sum of three angles of ΔAPN = 180]

= 90° - (90° - ∠BPN) [∵ ∠ANB = 90°, angle subtended by diagonal]

= ∠BPN

∠PNA = ∠BNP = 90°

By AA rule of similarity, ΔAPN ~ ΔPBN

$\frac{AN}{PN} = \frac{PN}{BN}$ [ratios of corresponding sides]

Or $PN^2 = AN \times BN = (10 + 5)(10 - 5) = 225$

∴ $PN = \sqrt{225} = 15\ unit$ [Ans]

3.16. This problem can be solved in multiple ways. The crux is to find the radius. The Origin O(0,0), point of tangency P and the centre C(5,0) form a right triangle with right angle at P.

Solution 1: Using linear graph approach
Find the equation of line through C and P in the slope intercept form $(y-0) = m(x-5)$
CP is perpendicular to y=2x, so m = -½
P is the intersection point of y=2x and
$(y-0) = -\frac{1}{2}(x-5)$
Comparing y, you get
$2x = -\frac{1}{2}(x-5)$ *or* $x = 1$, *and* $y = 2x = 2$. i.e P is located at (1,2)

$$\overline{CP} = \sqrt{(x_1 - x_2)^2 + (y_1 - y_2)^2}$$
$$= \sqrt{(5-1)^2 + (0-2)^2} = \sqrt{5}$$

∴ Area of the circle = $\pi(5)^2 = 5\pi$

Solution 2: Using geometric construction Drop PN ⊥ OC

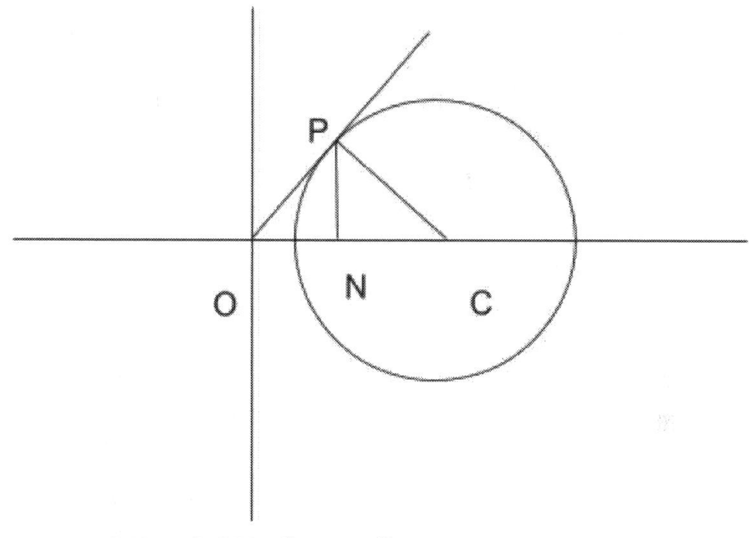

$PN = 2\,ON$, from y=2x

Now ΔOPN ~ ΔPCN, so ratio of corresponding sides

$\frac{CN}{PN} = \frac{PN}{ON}$

Or $CN = PN \times \frac{PN}{ON} = \frac{PN}{ON} \times \frac{PN}{ON} \times ON = 4\,ON$

But OC = 5

$\Rightarrow 5ON = 5 \Rightarrow ON = 1,\ PN = 2 \times ON = 2,\ CN = 4$

$PC = \sqrt{PN^2 + CN^2} = \sqrt{5}$

$\therefore Area\ of\ circle = \pi(\sqrt{5})^2 = 5\pi$

S4. Functions

1. Given: 3 and 10 are the zeroes of the quadratic function.
   So (x-3) and (x-10) are the factors of the function
   So $f(x) = a(x-3)(x-10)$, a is a real number
   We will need to find using the maximum value.
   Path: 1

   If you have prior knowledge that the graph of a quadratic polynomial is a parabola and symmetric, with the the vertex on the axis of symmetry. Also zeroes are symmetric about the axis. So the axis passes through x= (3+10)/2 = 6.5

   $f(5) = a(6.5-3)(6.5-10) = -a\frac{49}{4} = 5$ (given maximum value)

   $\Rightarrow a = -\frac{20}{49}$

   $f(x) = -\frac{20}{49}(x-3)(x-10) = -\frac{20}{49}x^2 + \frac{260}{49}x - \frac{600}{49}$

   Path: 2

   With no prior knowledge about parabola, you can work this out like the following:

   f(x) =a(x-3)(x-10)

   = a (x^2 -13x + 30)

   $= a(x^2 - 2x(\frac{13}{2}) + 30)$ ...write -13x as -2ab

$= a[x^2 - 2x(\frac{13}{2}) + (\frac{13}{2})^2 + 30 - (\frac{13}{2})^2]$

$= a(x - \frac{13}{2})^2 + a(30 - \frac{169}{4}) = a(x - \frac{13}{2})^2 + a(-\frac{49}{4})$

Now $(x - \frac{13}{2})^2 \geq 0$, so for a < 0, f(x)'s maximum value is $a(-\frac{49}{4})$

So $a(-\frac{49}{4}) = 5 \Rightarrow a = -\frac{20}{49}$

$f(x) = -\frac{20}{49}(x-3)(x-10) = -\frac{20}{49}x^2 + \frac{260}{49}x - \frac{600}{49}$

Path: 3

Let $f(x) = ax^2 + bx + c$

$Sum\ of\ roots = \frac{-b}{a} \Rightarrow \frac{-b}{a} = 10 + 13\ or\ b = -13a$

$Product\ of\ roots = \frac{c}{a} \Rightarrow \frac{c}{a} = 30 \Rightarrow c = 30a$

$Maximum\ value = \frac{4ac - b^2}{4a} \Rightarrow \frac{120a^2 - 169a^2}{4a} = 5$

$\Rightarrow a(-\frac{49}{4}) = 5 \Rightarrow a = -\frac{20}{49}$

$f(x) =$

$-\frac{20}{49}x^2 + (-13)(-\frac{20}{49})x + 30(-\frac{20}{49}) = -\frac{20}{49}x^2 + \frac{260}{49}x - \frac{600}{49}$

2. The cubic polynomial in standard form is

   $f(x) = ax^3 + bx^2 + cx + d$

   Given: it passes through (0, 0) and f(1) =1 , f(2)= 2 and f(3) = 4.

   Using (0,0), we get 0=d

   From other three conditions we get

   $a + b + c = 1 \ldots. (i)$

   $8a + 4b + 2c = 2 \quad \ldots.(ii)$

   $27a + 9b + 3c = 4 \quad \ldots.(iii)$

   Eliminate c from (i) and (ii):

   $2a + 2b + 2c = 2 \ldots(i) \times 2$

Subtracting $(i) \times 2$ from (ii) you get $6a + 2b = 0 \quad ...(iv)$

Eliminate c from (i) and (ii):

$(iii) - 3(i) \Rightarrow (27 - 3)a + (9 - 3)b = 1 \Rightarrow 26a + 6b = 1 \; ..(v)$

Eliminate b from (iv) and (v)

$(v) - 3(iv) \Rightarrow 26a - 18a = 1 \Rightarrow a = \frac{1}{8}$

Substituting a in (iv), $b = \frac{-6a}{2} = -3a = \frac{-3}{8}$

Substituting $a \; and \; b \; in \; (i)$, $c = 1 - a - b = \frac{5}{4}$

The polynomial is $f(x) = \frac{1}{8}x^3 - \frac{3}{8}x^2 + \frac{5}{4}x$

3. If f(x) = x^3 + 4x + 2

$\Rightarrow f(2y + 1) = (2y + 1)^3 + 4(2y + 1) + 2$

$= 8y^3 + 12y^2 + 12y + 1 + 8y + 4 + 2$

$i.e. f(2y + 1) - 8y^3 + 12y^2 + 20y + 3$

4. $f\left(\frac{3+x}{2}\right) = 3x$

Try to find the value of x in terms of t, such that

$\frac{3+x}{2} = t \; i.e. \; x = 2t - 3$

$f(t) = f\left(\frac{3+x}{2}\right)|_{x=2t-3} = 3(2t - 3) = 6t - 9$

$Or \; f(x) = 6x - 9$

5. Given $f\left(\frac{2+3x}{5}\right) = x^2 + 5x + 4$,

Find the value of x in term of a another variable X such that $\frac{2+3x}{5}$ =X i.e. $x = \frac{5X-2}{3}$

$f(X) = f(\frac{2+3x}{5})|_{x=\frac{5X-2}{3}} = (\frac{5X-2}{3})^2 + 5(\frac{5X-2}{3}) + 4$

$= \frac{25}{9}X^2 + \frac{-20+75}{9}X + \frac{4-30+36}{9} = \frac{25}{9}X^2 - \frac{55}{9}X + \frac{2}{9}$

X can be replaced with any variable as long as mapping remains the same

$f(x) = \frac{25}{9}x^2 - \frac{55}{9}x + \frac{2}{9}$

6. There is a nuance in the given function f(x) =$\sqrt{x^2}$ + x You might take the square root as a clue for restricting the domain non-negative values. Look carefully - you have a x^2 underneath √ , with that √ is always operating on a positive value irrespective x being +ve or -ve. So you don't have any restriction in domain. For $x \geq 0,\ f(x) = x + x = 2x,\ so\ f(x)$ maps to all non-negative real values
   For
   $x < 0,\ \sqrt{x^2} = -x,\ making\ it\ positive \Rightarrow f(x) = -x + x = 0$
   So domain { x | x is a real number}, and range
   $\{y \mid y\ is\ a\ non-negative\ real\ number\}$

7. Answers are tabularized, with D and R denoting domain and range respectively:

a) f(x)=3^{-x} D: all real numbers R:positive numbers	b) f(x) = $\frac{2}{\lvert x-3 \rvert}$ D: all real number except 3 R: all positive real numbers	c) f(x) = $-(x-4)^{-2}$ D:all real numbers except 4 R:all negative real numbers
d) y =|x-3| +	e)	f) f(x) =

4 D: All real numbers R: All real numbers ≥4	$f(x) = x^3 + x$ D: all real numbers except 0 R: all real numbers except 0	$(x-2)^2 + 6$ D: All real numbers R: all real numbers ≥6
g) $y = \sqrt{x-3}$ D: all real numbers ≥ 3 R:all real numbers ≥ 0	h) $y = \sqrt{\lvert x-3 \rvert}$ D: all real numbers R:all real numbers	i) $y = 3x^{\frac{1}{4}}$ D:All real numbers ≥ 0 R: all real numbers ≥ 0

8. Graph of a linear function is straight line, Equation of a straight line passing through (3, 4) and (7, -8) is given by:
   $(y-4) = (x-3)\frac{-8-4}{7-3} \Rightarrow (y-4) = -3(x-3) \Rightarrow y = -3x + 13$
   (slope intercept form)
   ∴ y intercept is 13 and slope = -3.

9. Y-intercept of a function is the value of the function at x=0 i.e. f(0)
   $y=f(x) = 3x^3 + 5x^2 + 3x + 5$
   $f(0) = 5$
   To find the zeros i.e. the values of x where f(x) is 0, you will need to factorize f(x)
   $f(x) = 3x^3 + 5x^2 + 3x + 5$

$= 3x^3 + 3x + 5x^2 + 5$
$= 3x(x^2 + 1) + 5(x^2+1)$
$=(3x+5)(x^2+1)$

$(x^2+1) \geq 1$, So f(x) = 0, when (3x+5)=0 i.e. at $x = -\frac{5}{3}$

10. Given f(n) = f(n-1) + n and f(1) = 1
A technique you can use is find difference between two successive terms and the differences.
f(n) - fn(n-1) = n
f(n-1) - f(n-2) = n-1 [replacing n = n-1]
..
f(2) - f(1) = 2
Adding above expressions you get
f(n) -f(n-1) + f(n-1) -f(n-2) +f(n-2) -...+f(2) -f(1) = n+(n-1)+...+3+2
Or f(n) -f(1) = n+(n-1)+..+2 = n(n+1)/2 -1 [using 1+2+..+n = n(n+1)/2]
Or f(n) -1 = n(n+1)/2 - 1
Or $f(n) = \frac{n(n+1)}{2}$

11. Recap that the standard form of arithmetic progression is. $a_n = a_1 + (n-1)d$
Given explicit form of the arithmetic progression is :
a_n = 5n -11 = 5n -5 -6 =5(n-1)-6
So a_1 = -6, and common difference d =-6

You can also do it as:

Get a_1 using n= 1, i.e. $a_1 = 5(1) -11 = -6$
d= $a_2 - a_1 = 5(2) -11 - (-6) = 10 -11 +6 = 5$

12. Try to write in the standard form $a_n = a_{n+1 -1} = 7(n +1) +11$ [using the given form]
$= 7n + 18 = 7(n-1) +25$
Comparing with the standard form, $a_1 = 25$ and d =7

13. y = f(x)+4 is vertical translation of y=f(x) by 4 units up. So new value would be at (4, -6+4) i.e. at (4, -2)
New y intercept would be 3+4 =7.

14. $y=-4(x-3)^2+2$ is translated version of $y=-4x^2$, with origin shifted to (3,2)
$y=-4(x-2)^2 -4$ is translated version of $y=-4x^2$, with origin shifted to (2,-4)

Graph II is Graph I translated 6 unit below and 1 unit to the left. So new maximum point is at (2, -4).

.

15. f(x) has zeros at x= 4, 7 and 9. f(x-3) is f(x) translated to the right by 3 units. So new zeroes are (4+3), (7+3) and (9+3) i.e. 7, 10 and 12.

16. An exponential function is of the form $f(x) = ab^x$ where a and b are real numbers.
From the table we get:

$$f(0) = 2 \Rightarrow a \cdot b^0 = 2 \Rightarrow a \cdot 1 = 2 \Rightarrow a = 2$$
$$f(1) = 6 \Rightarrow 3b^1 = 6 \Rightarrow b = 3$$
$$So\, f(x) = 2 \cdot 3^x$$
$$s = f(3) = 2(3^3) = 54$$
$$f(t) = \tfrac{2}{3} \Rightarrow 2(3^t) = \tfrac{2}{3} = 2(3^{-1}) \Rightarrow t = -1$$

17. arithmetic progression has has $a_{21} = 25$ and $a_{43} = 58$
Using standard explicit expression of arithmetic progression with common difference d,
$(a_{43} - a_{21}) = d(43\text{-}21) = d(22) = 58\text{-}25 = 33$
$\Rightarrow d = 33/22 = 1.5$
$\therefore a_1 = a_{21} - d(21\text{-}1) = 25 - 20(1.5) = -5$
Explicit formula for the progression: $a_n = -5 + 1.5(n\text{-}1)$

18. Rewrite the given line in slope intercept form : y = -(¾)x + 20/4. Slope = -3/4
Product of slopes of two perpendicular lines is -1
Slope of the required line = -1 / (-¾) = 4/3.
Given that it passes through (-10,2), its equation in point slope form:
$(y - 2) = (\tfrac{4}{3})\,(x - (-10)) \Rightarrow 3y - 4x - 46 = 0$

19. Given the sequence: $a_n = \frac{1}{4} - \frac{7}{20}(n - 2)$
Find a_1: $a_1 = \frac{1}{4} - \frac{7}{20}(1 - 2) = \frac{12}{20} = \frac{3}{5}$
Common difference :
$$a_n - a_{n-1} = -\tfrac{7}{20}(n - 2) + \tfrac{7}{20}(n - 3) = -\tfrac{7}{20}$$

The recursive definition is: $f(n) = f(n-1) - \frac{7}{20}$ $and\, f(1) = \frac{3}{5}$

20. Using $a_k = 25\; and\; a_{k-3} = 19 \Rightarrow a_k - a_{k-3} = 3d$, where d is the common difference.
    Or $6 = 3d \Rightarrow d = 2$
    Similarly $a_k - a_3 = d(k-3) = 2(k-3) = 2k - 6$, and
    $a_k - a_3 = 25 - 5 = 20$
    $Or\; 2k - 6 = 20 \Rightarrow k = 13$

21. Use explicit form of AP sequence: $a_n = a_1 + (n-1)d$
    $a_k + a_{n-k+1} = a_1 + (k-1)d + a_1 + (n-k+1-1)d$
    $= 2a_1 + (k-1+n-k+1-1)d = 2a_1 + (n-1)d$
    Rearranging we get:
    $a_k + a_{n-k+1} = 2a_1 + (n-1)d = a_1 + a_1 + (n-1) = a_1 + a_n$
    $S_n = a_1 + a_2 + a_3 + ... + a_{n-1} + a_n \; ...(i)$
    Writing S_n in the reverse order :
    $S_n = a_n + a_{n-1} + a_{n-2} + ... + a_2 + a_1 ....(ii)$
    Adding (i) and (ii), you get
    $2S_n =$
    $(a_1 + a_n) + (a_2 + a_{n-1}) + (a_3 + a_{n-2}) + ... + (a_{n-1} + a_2) + (a_n + a_1)$
    $= (a_1 + a_n) + (a_1 + a_n) + ... + (a_1 + a_n) + (a_1 + a_n)$ [n *times, using expression for* $a_k + a_{n-k+1}$]
    $= n(a_1 + a_n)$
    $\therefore S_n = \frac{n(a_1 + a_n)}{2}$ Ans

22. In the sequence 2, 9 , 16, 23, … common difference , d = 7

    Let 373 be the kth term

    $2 + 7(k-1) = 373$

    $\Rightarrow 7k - 5 = 373 \Rightarrow 7k = 378 \Rightarrow k = 378 \div 7 = 54$

    373 is the 54th term.

23. Let n be the 1st of the 11 consecutive integers.

    The last one in the series = n+10

    Sum of the series = 11 (1st + last)/2 =11(n + n+10)/2 = 11(n+5)

    Per given condition 11(n+5) = 121 or n+5 = 11 or n=6

    i.e Starting number is 6.

24. Let n be starting number of the 23 consecutive odd numbers.

    The last number of the series is n +(23-1)2 = n+44

    Sum of the series =

    $23(1st + last)/2 = 23(n + n+44)/2 = 23(n+22)$

    Given, sum of the series = 2047

    $\Rightarrow 23(n+22) = 2047 \Rightarrow n+22 = 89 \Rightarrow n = 67$

    The starting odd number is 67

25. RHS = $3^2 \cdot 3^4 \cdot 3^6 \cdot \ldots \cdot 3^{2n} = 3^{2+4+6+..+2n}$

    $= 3^{2(1+2+3+..+n)} = 3^{n(n+1)}$ [,, using 1+2+3+..+n = n(n+1)/2]

    LHS = $81^n = (3^4)n = 3^{4n}$

    Equating powers on both sides of the equation, you get

    $4n = n(n+1) \Rightarrow 4 = n+1$, *dividing both sides by n as* $n \neq 0$

$n = 4 - 1 = 3$

S5. Speed, Time and Work

5.1. Width of a pole is negligible compared to the length of the train. A train crosses a pole when its entire length has crossed the pole.

Speed = 60kmph = $\frac{600 \times 1000}{3600}\ ms^{-1} = \frac{500}{3} ms^{-1}$

So time taken to cross a pole = time taken traverse its own length = distance/speed

$$= 200 \div \frac{500}{3} = 1.2\ sec$$

5.2. Trains length =200m=0.2km, Length of bridge=1.3km

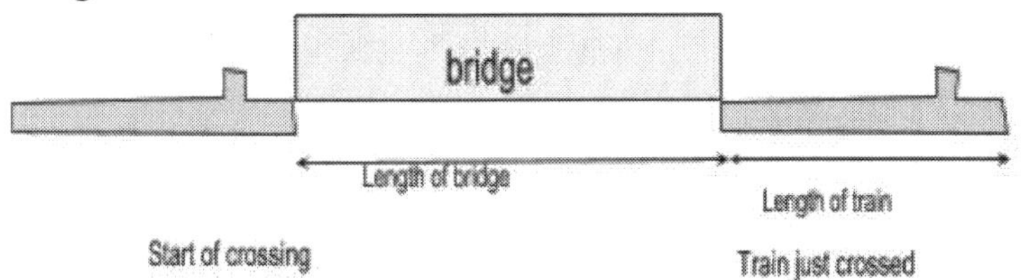

From the picture, train's head covered (length of bridge + its own length) distance

Time taken to cross the train = distance /speed =

$\frac{1.5km}{60kmph} = \frac{1.5}{60} \times 60\ min = 1.5\ min$

5.3. Anne goes ahead 25% of the track in Bobby's 1 round.

So she will go ahead 100% of the track in Bobby's 100/25 = 4 rounds.

So Bobby covers 4 rounds and Anne 5 rounds.

5.4. Natalie runs 30% faster than Pollie

I.e Natalie is 30% of track ahead. when Pollie finishes 1 round of the track

⇒ Natalie covers 1 full round extra, when Pollie covers 100/30=3⅓ rounds.

I.e. When they cross each other, they are ⅓ round away from the starting point.

The required shortest distance = 420(⅓) = 140m

5.5. Rita paints the wall in 12 days.

In 3 days, she paints (3/12) = ¼ of the wall.

Fraction of painting remaining = ¾

Rita's share of painting in next 6 days = 6/12 =½

Her daughter's share of painting in 6 days = ¾ -½ = ¼

Ie Her daughter can finish the painting alone in

$6 \div \frac{1}{4} = 24$ days.

5.6. A and B together can finish the work in 6 days.

⇨ A and B together finishes ⅙ th the work in 1 day.

I.e A's 1 day work + B's 1 day of work = ⅙ work

Note: In this kind of work-day problem , finish daily fraction of work

Similarly B's 1 day of work + C's 1 day of work = 1/9 work

And C's daily work + A's daily work = 1/12 work

By summing daily work fractions above, we get

2(A's, B's and C's daily work fractions) = ⅙ + 1/9 +1/12

$= \frac{6+4+3}{36} = \frac{13}{36}$

Sum of A's, B's and C's daily work fractions = $\frac{13}{72}$

A's daily work fraction = $\frac{13}{72} - \frac{1}{9} = \frac{5}{72}$

B's daily work fraction = $\frac{13}{72} - \frac{1}{12} = \frac{7}{72}$

C's daily work fraction = $\frac{13}{72} - \frac{1}{6} = \frac{1}{72}$

Recap, the inverse relation: if a person finishes x fraction of work in 1 day, he finishes the work in $\frac{1}{x}$ *days*

If they worked individually,

A finishes in $1 \div \frac{5}{72} = 72/5 = 14.4\ days$

B finishes in $1 \div \frac{7}{72} = 72/7 = 10\frac{2}{7}\ days$

C finishes in $1 \div \frac{1}{72} = 72\ \ days$

5.7. Assume tank's capacity be V litres

Filling rate with input = V/12 l/hr

Emptying rate with outlet = V/20 l/hr

When both inlet and outlet are open, overall filling rate = V/12 -V/20 = V/30 l/hr

Time taken fill remaining empty = Quantity/rate =(V/2) /(v/30) = 15 hrs.

5.8. Let V litres be the capacity of the tank.

Water filling rate with each faucet are V/10 l/min, v/15 l/min and V/20 l/min.

Amount of collected in 2 min when all three are open = (time)(overall rate)

= (2min) (V/10 + V/15 + V/20 l/min) = 2V($\frac{6+4+3}{60}$) = $26V/60\ l$

Remaining water volume = V - 26V/60 = 34V/60 l.

Combined rate when last two are open = V/15 + V/30 l/min = V/10 l/min

Required time to fill the remaining tank = amount/rate = (34V/60) /(V/10) = 34/6 = 5 $\frac{4}{6}min$

5.9. You can solve this problem purely algebraically, or you can do a bit reasoning and solve.

Algebraic Solution:

To solve this problem, you can assume the length of the tunnel to be L km and the distance of the train from the near end to be d km.

Using first boy just makes it: train traverses d and the boy L/3 in the same time.

i.e. $\frac{d}{speed\ of\ train} = time\ for\ the\ train = \ time\ for\ 1st\ boy\ = \frac{L/3}{v}$

$$\Rightarrow \frac{d}{60} = \frac{L}{3v} \Rightarrow d = \frac{20L}{v}\ ...(i)$$

Second boy also makes it, ie the train covers L+d and the boy ⅔ L in the same time.

I.e.

$\frac{d+L}{speed\ of\ train} = time\ for\ the\ train = \ time\ for\ 2nd\ boy\ = \frac{2L/3}{v}$

$$\Rightarrow \frac{d+L}{60} = \frac{2L}{3v} \Rightarrow d+L = \frac{40L}{v}..(ii)$$

Subtracting (i) from (ii) we get

$L\ = \frac{20L}{v} \Rightarrow v = 20\ i.e.\ v\ = \ 20kmph$

Arithmetic Solution:

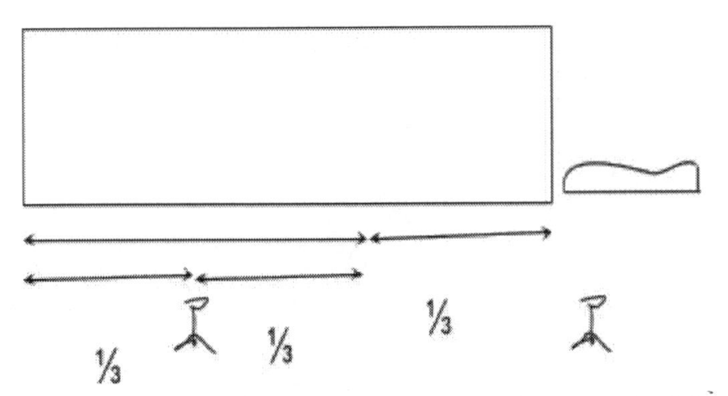

When boy at the nearer end reaches the mouth of the tunnel, the other boy also covers ⅓ tunnel. He will cover remaining ⅓ in the same time when the train will be covering the full tunnel.

So speed of the boy is ⅓ speed of the training = ⅓ (60) = 20 kmph.

5.10. The trains are approaching on same track with speeds 60 kmph and 90 kmph.

Their relative speed = 60kmph + 90kmph = 150 kmph = $150\ kmph$

Time taken to close the gap of 1.5 km is 1.5/150 = .01 hr.

Distance traversed by the bee in 0.01 hr = speed × time = 120 km /hr× 0.01 hr= 1.2 km

So total distance covered by the bee before getting crashed = 1.2 km.

5.11. Let the length of the trail be l km.

Time taken to cover l km during the forward journey = distance/speed = l/30 hr.

Time taken during the return journey = (l km) /(20 kmph) = l /20 hr.

During the to and fro journey, total distance covered = 2l km, and total time taken is

$\frac{l}{20} + \frac{l}{30}\ hr = \frac{5l}{60}\ hr = \frac{l}{12} hr$

∴ average speed = $\frac{total\ distance}{total\ time} = \frac{2l}{(l/12)} = 24\ kmph$

5.12. Although lengths of both the trains are given, JK sees the length of train B only.

Relative speed of JK wrt train B is 36kmph + 24 kmph = 60kmph = 1km/min.

So time noted by JK is same as time taken to cover the length of train B(200m =0.2km), at 1km/min is 0.2 min= 12 sec.

If the question asked for time taken to cross each other, you would need to add both train's length. The required time would be the time taken to cover the total length at the relative speed.

5.13. Note this problem has speeds specified in two units. All the instances of the same quantity, which is speed here, need to be converted into the same unit.

v_{JK}, Speed of JK, = 4 m/s

v_{MJ}, speed of MJ, = 7.2 km/hr = =

$7.2 \frac{km}{hr} \times \frac{1000m}{1km} \times \frac{1hr}{3600s} = 2m/s$

It's given implicitly that they are running in the same direction, as MJ was ahead of JK.

I.e. JK started from 200m behind MJ, and reached 100m ahead of MJ.
Relative to MJ, JK covers 300m.
Relative to MJ, JK's speed is 4m/s - 2m/s = 2m/s.
Time taken by JK = distance/speed =
$300m \div (2\ m/s) = 150s$

5.14. Duration between shots heard got reduced by 45 seconds. Ie. the man covered same distance in 20 min 15 seconds as is covered by sound in 45 seconds.

$\frac{speed\ of\ man}{speed\ of\ sound} = \frac{time\ taken\ by\ sound}{time\ taken\ by\ man} = \frac{45}{1215} = \frac{1}{27}$
Speed of the man = $\frac{1}{27} \times 351\ m/s = 13\ m/s$

5.15. You can solve this problem algebraically by assuming the length of each train or arithmetically using unitary method.

Algebraic Solution:
Assume length of each train to be x m.
Speed of trains are x/12 m/s and x/18 m/s.
Relative of the trains when they approach each other
$= \frac{x}{12} + \frac{x}{18}\ m/s = \frac{3x}{36} + \frac{2x}{36}\ m/s = \frac{5x}{36}\ m/s$
To cross each other two trains have to cover 2x m with this relative speed
Time taken to cross each other =
$2x\ m \div (\frac{5x}{36}\ m/s) = 14.4\ s$

5.16. Speed of boat A downstream = speed of A in still water + speed of stream= 9kmph

Speed of A upstream = speed of A in still water - speed of stream= 7kmph
Subtracting the 2nd expression from the first, you get 2(speed of stream) = 9-7 kmph = 2kmph
Speed of stream = 1 kmph.

Now downstream speed of slower boat is 8 kmph.
Its speed in still water = upstream speed - speed of stream = 7kmph.

Speed of A in still water = 9-1= 8kmph.
Time taken by A to cover 112km in still water = distance/speed = 112/8 = 14hr
Time taken by B to cover 112 km in stall water = 112/7 = 16hr.

The slower boat takes 2 hrs more.

5.17. Let the distance between two stations be x km.
During the forward journey, the train traverses first x/2 km at 300kmph, and the remaining x/2 km at 360kmph.
Total time taken for the forward journey =
$\frac{\frac{x}{2}}{300} + \frac{\frac{x}{2}}{360} hr = \frac{11x}{3600}\ hr$
During the return journey, the train covers 2x/5 km at 240 kmph and remaining 3x/5 km at 270kmph
Total time taken for the return journey =
$\frac{\frac{2x}{5}}{240} + \frac{\frac{3x}{5}}{270} hr = \frac{x}{600} + \frac{x}{450}\ hr = \frac{14x}{3600}\ hr$

So total time of the to and fro journey =

$\frac{11x}{3600} + \frac{14x}{3600}\,hr = \frac{25x}{3600}\,hr \;= \frac{x}{144}\,hr$

Total distance covered in the to and from journey = 2x km

∴ Average speed during the to and fro journey =

$\frac{total\ distance}{total\ time} = \frac{2x}{(x/144)}\ kmph \;=\; 288 kmph$

5.18. Ratio of speeds of April and May = 9:12 = 3:4.

Distance (covered in the same time) ∝ speed.

Ratio of distance covered by April and May = 3:4 i.e. April covers 3/7 of the track and May 4/7 of the track.

Ratio of the remaining tracks to be covered by April and May = 4/7 : 3/7 = 4:3.

Required ratio of speeds, if April meets May at the start point = 4:3

Required speed of April = (4/3) (May's speed) =

$\frac{4}{3}(12) \;=\; 16\ kmph$

Note: You can solve the problem algebraically assuming the length of the track, time when they meet and solving the equations so formed.

5.19. This problem can be solved both arithmetically using unitary approach and algebraically.

Arithmetic Solution:

Remaining work = ⅗ of total work, remaining time = ¼ scheduled time.

⅖ of work is finished in ¾ scheduled time, working 8hrs day, by 20 workers

∴ 1 work is finished in ¾ scheduled time, working 8hrs day, by $20 \times \frac{5}{2} = 50$ workers

∴ 1 work is finished in 1 scheduled time, working 8hrs day, by $50 \times \frac{3}{4}$ workers

∴ ⅗ work is finished in 1 scheduled time, working 8hrs day, by $50 \times \frac{3}{4} \times \frac{3}{5}$ workers

∴ ⅗ work is finished in ¼ scheduled time, working 8hrs day, by $50 \times \frac{3}{4} \times \frac{3}{5} \times \frac{4}{1} = 90$ workers

∴ ⅗ work is finished in ¼ scheduled time, working 10 hrs day, by $90 \times \frac{8}{10} = 72$ workers

Number of extra workers needed = 72-20 = 52.

5.20. The problem can be solved arithmetically using unitary approach or algebraically using man-day and woman-day approach.

Assume work 1 manday be m , and 1 woman-day be w, the work is n man-day.

Using 3 men and 5 women can finish a piece of work in $2\frac{2}{5}$ days

$3(2\frac{2}{5})\,m + 5(2\frac{2}{5})\,w = n\,m$ or

$\frac{36}{5}m + 12\,w = n\,m$ ....(i)

Using 6 men and 4 women can finish the same piece of work in $1\frac{7}{8}$ days,

$6(1\frac{7}{8})m + 4(1\frac{7}{8})\,w = n\,m$ or

$\frac{45}{4}m + \frac{15}{2}\,w = n\,m$ ...(ii)

Equating two equations, $\frac{36}{5}m + 12\,w = \frac{45}{4}m + \frac{15}{2}\,w$

$\Rightarrow \frac{24-15}{2}w = \frac{225-144}{20}m$

$\Rightarrow \frac{9}{2}w = \frac{81}{20}m \Rightarrow w = .9m$

Substituting in (i), $Total\ work = 7.2m + 10.8\ m = 18\ m$

Per day work of 9 men and 10 women = 9 m + 10 w = 18 m

9 men and 10 women can finish the work in 1 day.

S6. Probability

6.1 Let D be the event that the dart lands on the dart board and S be the event that dart lands in the shaded area.

P(S/D) = $\frac{Area\ of\ the\ shaded\ area}{Are\ of\ the\ dart\ board} = \frac{\pi a^2 - a^2}{\pi a^2} = (1 - \frac{1}{\pi})$,

assuming radius to be a

Given $P(\overline{D}) = 0.2$ So $P(D) = 1 - P(\overline{D})$ = 1 -0.2 = 0.8

∴ P(S) = P(D)* P(S/D) = $0.8(1 - \frac{1}{\pi})$

6.2 You can visualize the Teacher's set to be a container that contains 7 practiced set problems and 3 un-practiced. Three problems were picked from teacher's set random without replacement.

P(1st exam problem is from the practiced set) =

$\frac{Number\ of\ problems\ in\ Practice\ Set}{Number\ of\ problems\ in\ Teacher's\ test} = \frac{7}{10}$

After the first draw you are left with 6 problems in the Practiced set, and 9 problems in the teacher's test.

P(2nd exam problem is from the practiced set, Given 1st one was Practiced) = $\frac{6}{9}$

By the reasoning above, P(3nd exam problem was practiced Given 1st two were practiced) = $\frac{5}{8}$

Probability of all the three problems on the test were practiced by him

= $\frac{7}{10} \times \frac{6}{9} \times \frac{5}{8} = \frac{7}{24}$

6.3 Let AB be the event that drawn Ace is black

There are 4 aces and two aces are black

P(AB) = 2/4 = ½

$P(\overline{AB}) = 1 - \frac{1}{2} = \frac{1}{2}$

Given the drawn Ace is black

Number of black cards = 25,

Number of Jacks = 4

P(two cards are black or Jack)

= P(JJ) + P(BB) -P(both cards are jack and black)

$= \frac{4}{51} \times \frac{3}{50} + \frac{25}{51} \times \frac{24}{50} - \frac{2}{51} \times \frac{1}{50} = \frac{610}{2550} = \frac{61}{255}$

Given the drawn Ace is Not black

Number of black cards = 26, Number of Jacks = 4

P(two cards are black or Jack)

= P(JJ) + P(BB) -P(both cards are jack and black)

$= \frac{26}{51} \times \frac{25}{50} + \frac{4}{51} \times \frac{3}{50} - \frac{2}{51} \times \frac{1}{50} = \frac{660}{2550} = \frac{66}{255}$

Overall P(two cards are black or Jack) =

$P((BB\ or\ JJ)/AB)P(AB) + P((BB\ or\ JJ)/\overline{AB})P(\overline{AB})$

$= \frac{1}{2} \times \frac{61}{255} + \frac{1}{2} \times \frac{66}{255} = \frac{127}{610}$

6.4 Let W be the event that a visit to the casino results in a win.

P(W) = 0.2

L be the event that the visit results in a loss.

P(L) = 1- P(W) = .8.
Two wins in 4 visit corresponds to any of the sequences from
{WLLW,WLWL,WWLL,LLWW,LWLW,LWWL}

Probability of each sequence is same as there are two wins and two losses and equals (.2)(.2)(.8)(.8)= 0.0256

Hence probability of 2 wins in 4 visits = 6×0.0256 = 0.1536

6.5 The two events are complement of each other.
so x+y = 1

$$8x^2 + 16xy + 8y^2 + 10(x+y) = 8(x+y)^2 + 10(x+y)$$
$$= 8(1) + 10(1) = 18$$

6.6 Number of permutations possible with total 7 boys and girls = 7!
Number of permutations in which each end has a boy = (number of ways 1st end can be filled)(number of ways other end can be filled)(number of ways remaining 5 spots can be filled) = 4(3)5! = 12(5!).
Assuming each permutation to be equally likely,
P(at least one boy on each side in a photo) = 12(5!) /7!
= 12(5!)/(7×6×5!) =12/(7×6) = $\frac{2}{7}$

6.7 Number of sitting arrangements possible with six people = 6!

Number of arrangements where 3 friends are sitting together can be found by considering three friends as a single entity. So number permutations possible with 4 entities (entity of friends + three other people) = 4!
3 friends can permute amongst themselves in 3! Ways.
∴ Number of sitting arrangements in which three friends sit together = (3!) 4!
Assuming each sitting arrangement to be equally likely,
P(three friends find themselves sitting next to each other) = $\frac{(3!)\,4!}{6!} = \frac{(6)4!}{(6)(5)4!} = \frac{1}{5}$

6.8 There are 35 numbers in the set { 31, 32, .., 65}
Number of two number combinations possible = 35×34/2 = 35×17
Number combinations that add to 73 = {(36,37), (35,38),(34,39),(33,40),(32,41), (31, 42)}
Probability(both numbers add up to 73) = $\frac{6}{35\times37}$

6.9 The combinations needed for points to add up to 8 are {(2,6), (3,5),(4,4),(5,3),(6,2)}
P(2,6) = P(2)P(6)= $(\frac{1}{6})(\frac{1}{6}) = \frac{1}{36}$
Similarly probability of each combination is $\frac{1}{36}$, and there are 5 mutually exclusive combinations as listed above.
∴ P(points add up to 8) = $5(\frac{1}{36}) = \frac{5}{36}$

6.10 Let C_i be the event that a ball of color C which one of Red, blue and green is drawn from the urn U_i, i is one of 1, 2 and 3.

$P(R_1) = 2/7$, $P(B_1) = 3/7$, $P(G_1) = 2/7$
$P(R_2) = 3/8$, $P(B_2) = 2/8$, $P(G_2) = 3/8$
$P(R_3) = 1/3$, $P(B_3) = 1/3$, $P(G_3) = 1/3$

Drawing balls from any two urns are independent events.

P(drawing three different color balls) = $P(R_1B_2G_3) + P(R_1G_2B_3) + P(B_1R_2G_3) + P(B_1G_2B_3) + P(G_1B_2R_3) + P(G_1R_2B_3)$

$$=(2/7)(2/8)(1/3) + (2/7)(3/8)(1/3) + (3/7)(3/8)(1/3) + (3/7)(1/3)(3/8) + (2/7)(2/8)(1/3) + (2/7)(3/8)(1/3)$$

$$= \frac{14+6+9+9+4+6}{7\times8\times3} = \frac{48}{7\times24} = \frac{2}{7}$$

6.11 Given: Northbound trains are at 5:00 pm, 5:30 pm, and 6:00 pm.

Southbound(SB) trains are at 5:10 pm, 5:45 pm and 6:10 pm.

If Abbey arrives between 5:00-5:10 pm, he will see SB as the 1st train.

Again if he arrives between 5:30-5:45 pm, he will see SB as the 1st train.

He arrives randomly between 5:00-6:00 pm , i.e. 60 min interval

P(he sees SB as the 1st train) = P(he arrives in between 5:00-5:10pm)

+ P(he arrives in between 5:30-5:45pm)

$= \frac{10}{60} + \frac{15}{60} = \frac{25}{60} = \frac{5}{12}$

6.12 The set of numbers under consideration, S= {3, 5,7,11,13,17,19,23}

The pairs that add up to 24 are {(5,19),(7,17),(11,13)}

Number of equally likely pairs picked from S = Number of ways two numbers can be chosen from 8 uniques numbers= 8(8-1)/2 = 28

Number of favorable pairs = 3

∴ Probability(pair adds to 24) = $\frac{3}{28}$

6.13 The sequence must end in 24, 32, 36, 52,56,64

Let S_{ab} be the event that sequence ends in ab

$P(S_{ab})$ = P(1st draw is not a or b) P(2nd draw is a)P(3rd draw is b)

Now out of 5, 3 are not a or b

$\Rightarrow P(1st\ draw\ is\ not\ a\ or\ b) = \frac{3}{5}$

For the second draw, out of 4 in the urn, 1 is a.

$\Rightarrow P(2nd\ draw\ is\ a) = \frac{1}{4}$

For the 3rd draw, out of 3 in the urn 1 is b.

$\Rightarrow P(3rd\ draw\ is\ b) = \frac{1}{3}$

$\therefore P(S_{ab}) = \frac{3}{5}(\frac{1}{4})(\frac{1}{3}) = \frac{1}{20}$

P(Number is a multiple of 4)

$= P(S_{24}) + P(S_{32}) + P(S_{36}) + P(S_{52}) + P(S_{56}) + P(S_{64})$

$= 6(\frac{1}{20}) = \frac{3}{10}$

6.14 Let the length of side of the equilateral triangle be x.

Radius of the circle = (⅔)(altitude of the triangle) = (⅔)($\frac{\sqrt{3}}{2}a$) $= \frac{a}{\sqrt{3}}$

P(3) = $\frac{area\ of\ triangle}{area\ of\ circle} = \frac{\frac{\sqrt{3}}{4}a^2}{\pi(\frac{a}{\sqrt{3}})^2} = \frac{3\sqrt{3}}{4\pi}$

P(1) = $\frac{2}{3}(1 - \frac{3\sqrt{3}}{4\pi})$

$P(2) = \frac{1}{3}(1 - \frac{3\sqrt{3}}{4\pi})$

P(getting 4 with 2 darts)

= P(3)P(1) + P(1)P(3)+P(2)(2)

$= (\frac{3\sqrt{3}}{4\pi})\frac{2}{3}(1 - \frac{3\sqrt{3}}{4\pi}) + \frac{2}{3}(1 - \frac{3\sqrt{3}}{4\pi})(\frac{3\sqrt{3}}{4\pi}) + \frac{1}{3}(1 - \frac{3\sqrt{3}}{4\pi})\frac{1}{3}(1 - \frac{3\sqrt{3}}{4\pi})$

$= (1 - \frac{3\sqrt{3}}{4\pi})\{\frac{2}{3}(\frac{3\sqrt{3}}{4\pi}) + \frac{2}{3}(\frac{3\sqrt{3}}{4\pi}) + \frac{1}{3}(1 - \frac{3\sqrt{3}}{4\pi})\}$

$= (1 - \frac{3\sqrt{3}}{4\pi})\{\frac{\sqrt{3}}{\pi} + \frac{1}{3} - \frac{\sqrt{3}}{4\pi}\} = (1 - \frac{3\sqrt{3}}{4\pi})(\frac{1}{3} + \frac{3\sqrt{3}}{4\pi})$

6.15 Given 5 denominations are 2, 4, 6, 7 and 9 . Combinations of two denominations that add up to more than 13 are { (6,9) ,(7,7), (7,9), (9,7), (9,9) }

P(a combination of two draws with replacement) =

$(\frac{1}{5})(\frac{1}{5}) = \frac{1}{25}$

P(a total more than 13 in two draws) = $\frac{5}{25} = \frac{1}{5}$

6.16 Using, for two congruent polygons ratio of areas = square of scaling factor,

$$\frac{\textit{Area of inner triangle}}{\textit{Area of outer triangle}} = (\textit{scaling factor})^2 = x^2$$

Area of inner triangle= x^2(*area of dart board area*)

Total area of outer shell =

$(1 - x^2)$(*area of dart board area*)

Area of an inner lobe= $\frac{x^2}{3}$ dart board area

Area of an outer lobe= $\frac{1-x^2}{3}$ dart board area

P(hitting an even area)

$$= \frac{\textit{two inner sectors areas + one outer sector area}}{\textit{dart board area}}$$

$$= \frac{x^2}{3} + \frac{x^2}{3} + \frac{1-x^2}{3} = \frac{1+x^2}{3}$$

P(hitting an odd area) = $1 - \frac{1+x^2}{3} = \frac{2-x^2}{3}$

P(scoring odd in two hits)

= P(even in 1st hit)P(odd in 2nd hit) + P(odd in 1st hit)P(eve in 2nd hit)

$$= \frac{(1+x^2)}{3} \times \frac{(2-x)^2}{3} + \frac{(2-x^2)}{3} \times \frac{(1+x^2)}{3}$$

$$= \frac{2}{9}(1 + x^2)(2 - x^2) = \frac{2}{9}(2 + x^2 - x^4)$$

S7.Smallest and Largest Numbers

7.1 A number that is divisible by 10, 15 and 23 is a multiple of the LCM of these numbers.
LCM of 10, 15 and 23 is 690.
Smallest 4 digit number is 1000
Now $\frac{1000}{690} = 1\frac{310}{690}$ So the least 4 digit number divisible by 690 is 1000 + (690-310)= 1380.

Largest 4 digit number is 9999.

```
          14
690 | 9999
    | 690
    |-----
      3099
      2760
     -----
       339
```

Largest 4 digit number divisible by 690 is 9999 339 = 9660.
Required difference is 9660 - 1380 = 8280

7.2 Let's assume that there is a 4 digit perfect square number N that has more than one prime factors. For generality say x and y.

N is a perfect square, so each prime factor should be paired in the prime factorization of N.
I.e. $N = x^2y^2$
Now we assumed x > 60 and y >60, implying xy > 3600
So $N = x^2y^2 = (xy)^2 > (3600)^2 > 9999$, implying N cannot have 4 digits, contradicting our assumption.
I.e. A 4 digit perfect square cannot have more than one prime factors which are each greater than 60.
I.e. The number has only one prime factor that is greater than 60.
Now square root of a 4 digit number is a digit number.
So largest N will be given by the square of the largest 2 digit prime number.
Largest 2 digit prime number is 97, So the number is 97^2 i.e. 9409

7.3 Find the largest 3 digit number divisible by 93:
Largest 3 digit number is 999. Now 999/93 leaves a remainder of 69.
So largest 3 digit number divisible by 93 is 999 - 69 = 930
The number + 24 is divisible by 5
i.e. if the number is represent as XYZ, where X,Y and Z are the digits of the number,Z+4 should be either 5 or 10. $\Rightarrow$ *unit digit is* 1 *or* 6

930 = 93 $\times 10$, 93 $\times 9$ *ends in* 7, 93×8 *in* 4, 93×7 *in* 1.
The required number is $93 \times 7 + 24 = 651 + 24 = 675$

7.4 LCM and HCF of two numbers are 58 and 870.
Product of two numbers =LCM $\times$ HCF = 870 $\times$ 58 =
$58 \times 3 \times 5 \times 58$
The larger number is not divisible by the smaller number i.e. numbers are not (58, 870)
So the required number pair is
58×3 *and* 5×58 *i.e.* 174 *and* 290

7.5 Let r be the remainder in all cases and n be the number. So 78-r, 225-r and 414-r are all divisible by n.
Or (225-r) - (78-r) = 225-78=147 and
(414-r)-(225-r)=189 are divisible by n
Or greatest value of n is the HCF of 147 and 189,
HCF(147, 189) = 21
Required largest number is 21.

7.6 Three consecutive natural numbers can be assumed as (n-1), n and (n+1), where n >1.
Product of this tuplet is (n-1)n(n+1)
One of the three numbers is a multiple of 3, At least one of the factors is an even number.
6 is the largest number that will always divide the product of any three consecutive numbers.

7.7 Product of 4 consecutive natural numbers is of the form (n-1)n(n+1)(n+2)
Out of 4 numbers, 2 are even numbers, and out of 2 consecutive even numbers, one of them is a multiple of 4.

Also out of 4 integers, at least one is a multiple of 3,
The largest natural number that divides the product = $2 \times 4 \times 3$ =24

7.8 Try to find the square root of 9778 using the division method:

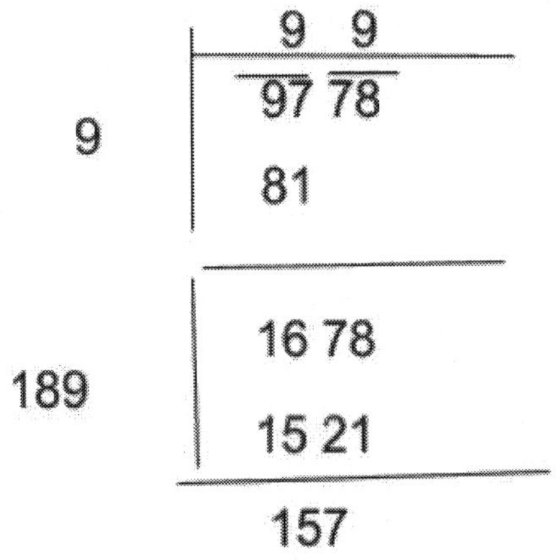

So $99^2 < 9778 < 100^2$
Required number of soldiers that must join the group = $100^2 - 9778 = 222$

7.9 If there are n number of full rows, total number of scouts needed for the formation = 1 + 2+..+n = $\frac{n(n+1)}{2}$
Find n such that $\frac{n(n+1)}{2} \geq 400$

$$\Rightarrow n(n+1) \geq 800 \Rightarrow n^2 + n + \tfrac{1}{4} \geq 800.35$$

$$\Rightarrow n + .5 \geq \sqrt{800} = 28.28 \Rightarrow n \geq 27.72$$

So with 400 scouts, 27 rows are completely full, 28 th is partially full.

Number of scouts in a 28 row formation with all rows full = $\frac{28(28+1)}{2} = 406$

Therefore number of scouts that need to stay back = $406 - 400 = 6$

7.10 Note the divisor -remainder differences:

10 -7 =3 ; 16 - 13 = 3 ; 24 -21 = 3

So remainder+3 is divisible by the divisor in each case.

So the dividend is 3 less the number which is divisible by 10, 16 and 24.

LCM(10,16,24) = $2 \times 5 \times 8 \times 3 = 240$

The required number is 240-3 = 237.

APPENDIX

A1. Exponent Rules

A1.1 The expression a^m is a power where a is the base and m is the exponent. Example 3^2 is a power of 3. Also $3^2 = 3 \times 3$.

A1.2 Product of power property:
$a^m \times a^n = a^{m+n}$, for example $3^2 \times 3^5 = 3^8$

A1.3 Quotient of power property:
$\frac{a^m}{a^n} = a^{m-n},\ a \neq 0$, for example $5^4 \div 5^2 = 5^2$

A1.4 Negative power property:
$a^{-1} = \frac{1}{a},\ a \neq 0,\ for\ example\ 3^{-1} = \frac{1}{3}$

A1.5 Zero power property
$a^0 = 1.\ Follows\ from\ 1 = \frac{a}{a} = a^{1-1} = a^0.\ Example\ 7^0 = 1$
$(x^2y^3)^0 = 1$

A1.6 Power of a power property:
$(a^m)^n = a^{mn}$ example $(3^2)^3 = 3^6$

A1.7 Power of a product
$(ab)^m = a^m b^m,\ example\ 6^3 = 2^3 \times 3^3$

A1.8 Power of a quotient
$(\frac{a}{b})^m = \frac{a^m}{b^m},\ example\ (\frac{3}{5})^3 = \frac{3^3}{5^3}$

A1.9 Radical is an expression of the form $\sqrt[n]{a}$. b is the nth root of a, if b^n =a. This is also expressed as $b = \sqrt[n]{a}$. For example, that 5 is the cube root of 125 can be written as $5 = \sqrt[3]{125}$

A1.10 Fractional exponent: n the root can also be expressed as $a^{\frac{1}{n}} = \sqrt[n]{a}$. In general is fractional exponent can be expressed as $a^{\frac{p}{q}} = \sqrt[q]{a^p} = (\sqrt[q]{a})^p$

A2. Commonly Used Formulae

A2.1 $(a+b)^2 = a^2 + 2ab + b^2$

A2.2 $(a-b)^2 = a^2 - 2ab + b^2$

A2.3 $a^2 - b^2 = = (a - b)(a + b)$

A2.4 $a^3 - b^3 = = (a - b)(a^2 + ab + b^2)$

A2.5 $a^3 + b^3 = = (a + b)(a^2 - ab + b^2)$

A2.6 $(a+b)^3 = a^3 + 3a^2b + 3ab^2 + b^3$. $(a-b)^3$ can be found by replacing b with -b.

A2.7 Sum of first n natural number, i.e 1+2+ ..+n = $\frac{n(n+1)}{2}$

A2.8 Sum of the squares of first n natural numbers:
$1^2 + 2^2 + 3^2 + ... + n^2 = \frac{n(n+1)(2n+1)}{6}$

A2.9 Arithmetic Progression: A sequence where difference between two consecutive terms is same. For example 2,6, 10, 14, ... is an arithmetic sequence. N th term is related to the first term as: $a_n = a_1 + (n-1)d$, where d is the difference.

A2.10 Sum of an arithmetic sequence with n terms = $n \times \frac{a_1 + a_n}{2}$

A2.11 Geometric Progression: A sequence of the form a, ar, ar^2,ar^3,.. Is a geometric progression. Two consecutive terms have the same ratio. $a_{i+1} = ra_i$

A2.12 Sum of a geometric sequence with n terms = $a_1 \frac{r^n - 1}{r - 1}$

Made in the USA
San Bernardino, CA
12 May 2019